SUPPLEMENTAL WAYS FOR IMPROVING INTERNATIONAL STABILITY 1998

(SWIIS'98)

A Proceedings volume from the IFAC Conference,
Sinaia, Romania, 14 - 16 May 1998

Edited by

I. DUMITRACHE
Control and Computer Faculty, 'Politehnica' University of Bucharest,
Bucharest, Romania

F. KILE
Microtrend, Appleton, Wisconsin, USA

and

P. KOPACEK
Institute for Handling Devices and Robotics,
Vienna University of Technology,
Vienna, Austria

Published for the

INTERNATIONAL FEDERATION OF AUTOMATIC CONTROL

by

PERGAMON
An Imprint of Elsevier Science

UK	Elsevier Science Ltd, The Boulevard, Langford Lane, Kidlington, Oxford, OX5 1GB, UK
USA	Elsevier Science Inc., 660 White Plains Road, Tarrytown, New York 10591-5153, USA
JAPAN	Elsevier Science Japan, Tsunashima Building Annex, 3-20-12 Yushima, Bunkyo-ku, Tokyo 113, Japan

First edition 1998

Library of Congress Cataloging in Publication Data

A catalogue record for this book is available from the Library of Congress

British Library Cataloguing in Publication Data

A catalogue record for this book is available from the British Library

ISBN 0-08-043231 X

Transferred to Digital Printing 2009

IFAC CONFERENCE ON SUPPLEMENTAL WAYS FOR IMPROVING INTERNATIONAL STABILITY

Sponsored by
International Federation of Automatic Control (IFAC)
IFAC Technical Committee on Supplemental Ways for Improving International Stability (SMW)

Organized by
Romanian Society for Control and Technical Informatics (SRAIT)

Co-sponsored by
IFAC Technical Committees on
- Cultural Aspects of Automation (HAR)
- Man-Machine Systems (SMM)
- Modelling and Control of National Economies (SME)
- Social Impacts of Automation (HAS)
- Automation in Developing Countries (IDE)

Romanian Ministry of Research and Technology (MCT)
Media-Tech S.A.
Felix Telecom

International Programme Committee (IPC)
F. Kile (USA) (Chairman)
C. Antonini (RSA)
H. Chestnut (USA)
Gh. Dimirowsky (MAC)
I. Dumitrache (RO)
P. Fleißner (A)
P. Groumpos (GR)
M. Hersh (UK)
J. Holubiec (PL)
K.H. Kienitz (BRA)
P. Kopacek (A)
St. Mittnik (GER)
A. Onishi (J)
V. Rasvan (RO)
J. Richardson (FR)
J. Scheffran (GER)
J. Scrimgeour (CDN)
A. Skorokhodov (RUS)
T. Vamos (H)
E. Welfonder (GER)

National Organizing Committee (NOC)
I. Dumitrache (Chairman)
Th. Borangiu
M. Arhip
S. Iliescu
M. Dragoicea
Th. Dragomir
Al. Marinescu
Gh. Tucu

FOREWORD:

For the past years, a number of people actively involved in engineering, but also in other fields, have worked on the topic of international stability. Workshops were held in various places on our world, and organized from different persons and institutions.

This conference, the sixth event in the SWIIS series, continued the tradition set in the earlier five SWIIS meetings. The goal was the beneficial application of systems engineering methods onto description of conditions, in which nations or groups interact with one another. Scientists from other fields such as political science, economics, social science, and international studies should have a platform to present and discuss their ideas.

In the technical program of this conference, 2 Survey Papers and 19 regular papers - grouped into 8 sessions - were presented to the very interested audience. Papers were given in the following areas:

- ⇨ Methodological analysis
- ⇨ Investigation of development: stability, sustainable development
- ⇨ Modelling of stability
- ⇨ Application of control principles to international stability
- ⇨ International policy co-operation
- ⇨ Cultural and educational aspects in international stability
- ⇨ East/West/North/South relationships
- ⇨ Global development – Regional impact
- ⇨ Negotiation and Mediation in Conflict

The many positive comments made by the participants of this conference may confirm its success in meeting its objectives.

Bucharest, Appleton
June 1998

Ion Dumitrache
NOC Chairperson

Fred Kile
IPC Chairperson

CONTENTS

SURVEY PAPERS

Sustainability, Peace, and Future Leadership 1
F. KILE

Political, Economic and Military Evolution of Instability, North Korea, 1945-1998 9
J.G. RICHARDSON

STABILITY

Virtual Enterprise - A Strategical Approach to Improve Socio-Economic Stability 15
I. DUMITRACHE, A.M. STANESCU, S.I. CARAMIHAI, A. CURAJ

Social Entropy, Technological Change and International Stability 21
S. DOBRESCU, R. DOBRESCU

Stability of Ecological Economic Linear Models (Sensitivity Analysis) 29
A. MIRONYCHEV

A Systems Approach to Understanding the Causes of Instability in Nations: A Case Study 35
M.A. HERSH

MODELLING

A Dynamic-Game Model of Cooperation in Energy and Climate Change 41
S. PICKL, J. SCHEFFRAN

Object Oriented Model to Support Early Warning and Rapid Reaction Planning in International Context 45
V.M. SHALAMANOV, T.D. TAGAREV

A Behavior Model for Preventive Health Practices 53
G. SIMION

Acts of Mathematics with Hierarchical Haze 57
S. NOVIKAVA, A. ZHYBUL

Artificial Societies: A New Paradigm for Complex Systems' Modelling 63
M. DASCALU, E. FRANTI, G. STEFAN

METHODOLOGIES

Stability Versus Uncertainty 69
V. IONESCU

Geopolitical Relations in Post USSR Europe as a Subject of Mathematical Modelling and Control 73
S. LEVKOV, A. MAKARENKO

Demographical Units in Hierarchical Coordinates: Construction, Activity and Goals 79
W. NOWIK, V. BARKALIN, W. ZIANEVICH, S. GANCHAROVA, P. BUKA, K. MIATLIUK, A. ASTROWSKAYA, A. BAHUTSKI

Art in State Constructing Process 85
A. BURAWKIN, S. GANCHAROVA, A. DANICHAW, S. IOSKA, A. TUROVICH, I. PADYNINOGINA, A ZEZULA, W. ZIANEVICH, A. ZHYBUL, Y. BLUDAW, H. OSIPAVA, Y. TERAHAW, T. ARESHKA, A. ASTROWSKAYA, D. SAROKINA, A. BLUDAW, I. YUHNEVICH

Practical Instance of State Design in Reconstructing Region 91
P. BUKA, A. BURAWKIN, W. NOWIK, Yu. PACIUPA, I. KUZMINICH

East-West Academic Cooperation within the Framework of TEMPUS-PHARE Program 97
O. PASTRAVANU, M. VOICU

For Higher Flexibility on the Labour Market 103
M. KALAIDJIEVA

INTERNATIONAL POLICY COOPERATION

Gains from International Macroeconomic Policy Coordination Under Alternative Objective Functions 107
G. HABER, W.J. McKIBBIN, R. NECK

Sway of New Time: General Lines 115
F.O. KILE, P.P. GROUMPOS, F.G. FILIP, S. NOVIKAVA

Author Index 121

SUSTAINABILITY, PEACE, AND FUTURE LEADERSHIP

Frederick Kile

Microtrend
232 Peppercorn Drive, Appleton,
WI 54915-7137, USA
e-mail: 102610.2345@CompuServe.COM

Abstract: Segments of human society have competed with one another while seeking growth and while relying on natural biosystems to support human life. Expansion of competing thought systems often results in war and instability. A variety of earlier attempts at stable peace have notendured -- because competing goal sets have been chosen. The paper argues that a single. coherent goal set is unlikely to emerge and that society should seek a new paradigm for peace with sustainability -- social and systemic goal preferences which accommodate variety. The paper further argues that past criteria for leadership do not permit the needed accommodation. The paper suggests criteria for future leadership which will succeed in overcoming Cassandra's Dilemma (the mythical prophetess, Cassandra, received a gift of perfect prophecy which would not be believed). By way of parallel, past warnings of social catastrophe have been ignored. not because those warnings were incorrect, but because they were not believed. New criteria for future leadership incorporate the orientation for success in a set of systems with limited future choice. *Copyright © 1998 IFAC*

Keywords: Sustainability, Leadership

1. INTRODUCTION

Most major philosophies and religions (together with their leaders) focus on one of two orientations toward society:

1. A process orientation -- improving current physical and social conditions.
2. A goal orientation -- achieving a defined goal -- political, economic, social.

In recent decades a third consideration has been unfolding -- sustainability of the natural biosystems which support human life. and ultimately society. Serious questions have emerged regarding limits to stress on these biosystems. The book, "Limits to Growth," and an earlier, more poetic counterpart, Rachel Carson's "Silent Spring," highlighted these serious questions.

The human desire to rise to positions of leadership has required aspiring leaders to reject notions of limits. Libraries are filled with political, business, philosophical, and military works on growth -- of empires, of sales and profits, of social progress, of conquest. These writings and their authors have presupposed that tomorrow will not only be better but will have more of whatever is deemed worth having, whether political adherents, clients and profits, believers in "this" philosophy or religion, or territory and populations which "we" rule. The underlying presupposition always assumes that the natural biosystems can support these aspirations.

These leaders assume unlimited resiliency of supporting biosystems; thus, to suggest fragility of underlying biosystems is to undermine a leader's philosophy. How do we build a sense of urgency?

What will cause people to believe that this era is different from prior eras?

In this paper, we are limited to rhetoric, but future leaders will find ways to persuade people that the time for action is now. It is difficult to document because of the highly-charged political character of the recent past, but there is substantial evidence that the recasting of the USSR into the Commonwealth of Independent States was due in part to the realization that the environment was collapsing in certain key areas of the former USSR. Acid rain, saline soils, air and water pollution, and finally the fallout from the Chernobyl disaster convinced many leaders that "dominance" over nature was costing too much, that change was needed. Some of the price of this unsuccessful attempt at dominance was paid because of an equally futile attempt to dominate other forms of social organization.

At present, growing segments of social leadership are recognizing that the disastrous decline of the world's fisheries is too costly to continue. In parallel, there is increasing concern about the threats of weapons of mass destruction. Leaders, together with news media, are gradually educating the broader public to the proliferation of both means of destruction and the increasing availability of these terrible weapons to smaller and smaller fringe groups. Perhaps these proliferations account for the increase in the number of "rogue leaders" or "rogue nations." Before weapons of mass destruction proliferated, "rogue leaders" were frequently in such militarily weak positions that they were simply ignored. Today, rogue leaders command attention because they are able to acquire weapons of mass destruction and because they seem willing to use these weapons. The new leadership paradigm will be challenged to deal with fringe elements of global society.

The future of supporting biosystems is of uncertain outcome due to stress on these systems, given the expansionist thrust of human aspirations and the poorly understood resiliency of the biosystems -- recognizing the interconnectedness of the social world, both among social systems and with the natural world (Novikava, *et al.*, 298).

Maintaining the health of our supporting biosystems is particularly challenging because needed actions seem to call for sacrifice in the present in return for uncertain gain in the future. This equates to paying a short-term price for a long-term gain with some uncertainty about whether there ever will be a gain. Some advocates of continued economic growth assume that technology can "fix" the environment if ever it is overstressed. These people believe that technology will provide answers if/when answers are needed. This brings us to our title: "Sustainability, Peace, and Future Leadership." Cassandra, a legendary Greek prophetess was condemned by a curse from one of the Olympian gods: Her prophecies would always be correct but never be believed. Building peace with sustainability of supporting biosystems requires that leaders adopt a new mindset -- a new orientation toward what is achievable. As thought systems compete with one another and vie for growth, several different outcomes are possible:

Possible Outcomes:

1. Struggles continues and peace remains elusive -- "Repeated Wars"
2. A tightly kit belief system supersedes prior systems of thought -- "Universal Philosophy"
3. Supporting biosystems weaken and social compression ensues -- "Social Collapse"
4. Existing systems of thought accommodate one another and supporting biosystems -- society reaches peace and sustainability -- "Accommodation"

Some experts and leaders have tried a variety of "technical fixes." Several have been implemented in the past. No technical fix has been successful at maintaining stable peace. Moreover, in spite of attempts at technical fixes to maintain a stable environment, the global environment has less "headroom" each decade. "Headroom" is the capacity of the environment to absorb pollution or some other exogenous shock (for example, a major volcanic event which alters the earth's surface temperature for several years) without suffering permanent deformation.

Unsuccessful technical fixes:

1. The League of Nations;
2. The United Nations;
3. Various "balance of power" arrangements;
4. The Green Revolution -- to provide adequate food supplies (population growth is catching up with food supplies)
5. "Elimination" of certain diseases through antibiotics. Subsequently, some bacteria have become resistant to antibiotics and diseases are reappearing; among them: syphilis, gonorrhea, staphylococcus infection, TB.

No earlier technical fix has succeeded over the long run. It seems doubtful that any technical fix will emerge to forestall one of the four outcomes suggested above:

1. Repeated Wars
2. Universal Philosophy
3. Social Collapse
4. Accommodation

This paper argues that outcomes 1 and 2 (Repeated Wars and Universal Philosophy) have been attempted frequently and unsuccessfully and that one of outcomes 3 or 4 is unavoidable, i.e., that society will collapse unless groups accommodate to differences. This challenge may be so distasteful to leaders that they will continue to ignore very clear warning signals from within society and to overlook apparently irreversible changes in supporting biosystems. Modern leaders are challenged to rise above Cassandra's Dilemma and persuade society to change throughaccommodation.

2. DISCUSSION OF POSSIBLE OUTCOMES

2.1 Repeated Wars

Certainly repeated wars are a possible outcome -- but this is an outcome that few people or nations want and it is an outcome which may lead to human annihilation. Thus, other possible outcomes are more desirable -- both in a relative sense and in an absolute sense.

2.2 Universal Philosophy

Numerous attempts have been made to establish a universal philosophy. Following the collapse of the Roman Empire, the Christian Church emerged in the Western and the Mideastern Worlds as dominant -- a form of universal philosophy. Buddhism dominated large portions of Asia in similar fashion. Over time these "universal" dominations eroded. In Asia, Hinduism re-emerged, Confucianism held its own, and ultimately Islam took a place alongside other religions (philosophies) as one among several. In North Africa and Europe, Islam competed for the role of "universal philosophy." As a more secular age replaced religious movements, various systems of thought competed for attention, but little progress was made toward anything universal. The Communist movement seemed on a path to domination and emergence as a possible "world system," but it, too, faded. At best, competing world systems lacked energy for all-out conflict. At worst, wars erupted as various thought systems and belief systems struggled one with another.

The historical record suggests that no universal philosophy can ensure global peace and stability.

2.3 Social Collapse

Given that humanity avoids repeated wars and that no universal philosophy emerges to gain worldwide allegiance, we can expect that social structures will continue on the current path: avoiding major wars and expanding by building and growing within current political and ethnic domains. Growth is possible through further "conquest of nature," i.e., by using more of the resources available and by expanding human activity into areas formerly inhospitable to humans.

At some point, supporting biosystems will weaken and degrade and the capacity of those systems to support human activity will diminish. There is already evidence that some regional systems have weakened and that their capacity to produce food or usable materials has been diminished. One of the most salient examples is the drying up of the Aral Sea and the disappearance of the regional fishing industry which once flourished and supported a significant population. As weakening of biosystems occurs elsewhere, human activity will also be curtailed. Overfishing of some species of North Atlantic fish is beginning to be felt in the nations which relied on these fisheries for major portions of their diet.

If major biosystems weaken, lose resiliency and contract, social compression will occur. "Social compression" begins when the supporting biosystems contract or when some other contraction occurs (loss of significant energy supplies, loss of territory, unfavorable climate change) in the larger domain within which a society operates. When social compression occurs, conflict begins and social structures break down. Ultimately the society collapses. Many "dead civilizations" died as a result of social compression. Other significant societies simply stagnated at a level consistent with the capacity of supporting biosystems to support human activity.

Some observers argue that human activity can "fix" biosystems when they weaken. In some instances natural systems have been "repaired" or "saved." Most of these "repaired" systems have been relieved of some major stresses before they collapsed. They then continued functioning at some reduced level of activity. Examples include:

1. The irrigated cotton lands of the Texas Panhandle. These areas were initially farmed as dry lands. With irrigation, the lands produced larger crops. As aquifers sank or as soils became saturated with unwanted minerals, some lands were returned to dry-land farming techniques. A return to traditional farming methods reduced the economic value of the crops, but it stabilized the land. Clearly, the land supported fewer people under dry-land farming than with irrigation.

2. Limiting catches of certain fish species (including the North Atlantic Cod) to avoid extinction of those species. When catch limits are imposed, the immediate impact is that some fishermen are forced out of the business. The longer-term impact is a potentially increased catch in the future.

The examples of dry land farming and commercial fishing illustrate some of the stresses experienced when biosystems are stretched near their limits. These two examples also indicate that there is some potential for "repairing" or "restoring" natural systems. The examples do not suggest what might happen if a number of systems supporting the same human population were all to need restoration at approximately the same time. In that instance, social compression would be experienced. Social compression places great stress on society, and at some level of social compression, society would erupt in violence and perhaps collapse altogether.

2.4 *Accommodation*

We are suggesting that "accommodation" is one (perhaps the only) mode of social behavior which is sustainable and which can undergird stable peace. At the very least, successful accommodation will require new leadership. New leadership will be required because prior generations of leaders and prior "leadership paradigms" have all assumed that the group being led will either triumph over all other groups or, at a minimum, join with other groups in subduing nature and physical limitations to "grow and conquer." It would be difficult toidentify any significant group of past leaders or past social groups which have not followed the paradigm of "grow and/or conquer."

Thus, in order to implement a strategy of accommodation successfully, new leadership paradigms must be developed. It will, of course, be important in the new leadership paradigm if leaders can amplify the economic interdependence which already exists. On the other hand, it is equally important that competitive rivalries not deteriorate to the point of hostility. Creating an image of a workable accommodation with each other and with our supporting biosystems will challenge future leaders.

In order to develop these paradigms, society must first be convinced that the path of accommodation is the most reasonable path to a sustainable future.

Accommodation has at least two defining characteristics:

1. Each social unit (at every level of aggregation) accommodates to the need for other social units to exist.

2. Each social unit recognizes that all social units (from the individual to global society) depend on shared biosystems to support human life, and that if these biosystems weaken, one (or more) of three outcomes will occur: Repeated Wars, A Universal Philosophy, Social Collapse.

Given that no philosophy has succeeded in becoming universal, one (or both) of the two remaining outcomes seems likely: War or Collapse. The goal of Accommodation cited in this paper could mistakenly be thought of as a Universal Philosophy, and thus beyond reach for humanity. Our modern dilemma might then be merely a re-living of Cassandra's Dilemma, i.e., contemporary society realizes what it must do to survive, but society does not understand its predicament sufficiently well to deal effectively with the predicament. As suggested below, accommodation does not purport to be a universal philosophy. It is rather a rubric intended to encourage and to empower varying philosophies or religions to exist side-by-side.

The goal of this paper is twofold:

1. To further clarify (what has already been stated by many observers) that human social systems must redesign themselves for survival.

2. To set forth a paradigm for a new style of leadership which will be able to motivate our various social units to cooperate for survival. We have no special knowledge of prior efforts in this direction (beyond simple dicta such as "think globally, act locally") and thus our attempts at setting forth a new paradigm are de novo and almost certainly flawed. We are pleased to offer an outline sketch of the new leadership paradigm, hoping that others will further develop this rudimentary model.

3. SUSTAINABILITY

We noted above that past philosophies as well as extant philosophies have had either a process orientation or a goal orientation. For consistency with contemporary terminology, we will place the new leadership paradigm under the rubric "Sustainability". The new paradigm will focus both on process and on a goal. The goal of the new paradigm is "continued stability" (steady state operation) -- in contrast with a "final" (static) system state. Thus, it is closer to a process orientation than to a goal orientation.

Our model is designed around shared accountability. Any alternative paradigm could easily be construed as "world government," which is anathema to most social groups.

4. NEW LEADERSHIP -- A BOLD PARADIGM

Shared accountability -- a definition: Accountability is shared among groups when no one group seeks dominance over any other group(s) and all groups, individually and collectively, act to foster sustainability of the biosystems on which the groups subsist.

The new leadership paradigm is offered under the assumption that all prior groups (and their leaders) acted in a destabilizing manner by seeking dominance over other groups or by bending -- and perhaps exceeding the sustainable limits of supporting biosystems. The new leadership paradigm recognizes that past leadership models have relied on methods which are highly non-linear: when war has been used as an instrument of policy, the old paradigms generally sought to win at all costs (under an assumption that defeat was costlier than winning); alternatively, when peace was attained, the generally dominant paradigm was governed by an unwillingness to change -- a "don't rock the boat" mentality. Thus, actions to perpetuate war had a high policy multiplier while policies to preserve peace could often be characterized as non-actions, i.e., having a low policy multiplier. Likewise, resilience of supporting biosystems was influenced by non-linear policy paradigms, characterized roughly as follows: before serious deterioration of supporting biosystems became evident, "do little or nothing" -- equivalent to a low policy multiplier and after serious deterioration of biosystems was detected, act radically -- occasionally through conservation efforts, but often through abandonment of the system: migration, seeking alternative supporting biosystems (witness: abandoned mines [not biosystems, but resources nevertheless], abandoned oil fields [as with mines, resources], abandoned farms, desertification, the closing of some North Atlantic fishing grounds by sovereign nations in hopes that decimated fish species will replenish their numbers); it should be noted that some of the largest migrations in history have taken place in recent years. These migrations have been due to the failure of accommodation -- either in politics or in a failure to accommodate supporting biosystems. Often, both factors are present in a migration. Mass migrations are truly radical actions.

As noted above, two common outcomes of past policy paradigms were:

1. War -- a costly, and eventually unsustainable, condition.
2. Loss of resilience in supporting biosystems -- ultimately an unsustainable course.

Goal: A paradigm for shared accountability assumes that future social sustainability requires a cessation of wars AND assumes that people and groups act within the stability limits of biosystems which support society. The new leadership must create a bold new paradigm because prior leadership models have proven unable to attain the goal set forth. In developing a new paradigm for leadership, two preliminary steps will be important:

1. Listing representative historically important modes of leadership;
2. Indicating how a new mode of leadership should transcend earlier modes of leadership to deal effectively with problems confronting humanity at the beginning of the third millennium A.D.

These preliminary steps can suggest needed characteristics of a new mode of leadership. At the same time, a model of future leadership should be simple and defined in positive terms. We will avoid defining leadership in terms of what it is not.

Some historically important modes of leadership

1. Authoritarian
2. Charismatic
3. Bureaucratic
4. Exemplary (role-modeling)
5. Communitarian
6. Technocratic
7. Mediating (building on enlightened self-interest)
8. Visionary

This list may not be exhaustive. Moreover, no list could be complete since new combinations of leadership attributes could define new modes of leadership.

The modes of leadership listed above are described briefly below. We have named people who exhibited characteristics of these modes of leadership, specifically omitting several religious leaders thought by many of their followers to be divine or quasi-divine: Jesus, Buddha, Mohammed. Many of the leaders we cite as examples exhibited characteristics of more than one mode of leadership. We simply cite these leaders as salient examples of particular modes of leadership.

1. Authoritarian -- wielding absolute power as a dictator, king, emperor (Caesar, Stalin, Saddam Hussein)

2. Charismatic -- drawing a following by force of personal character, charm, magnetism, or persuasiveness (Eva Peron, Martin Luther King)

3. Bureaucratic -- leading through the power of office, often elective or attained through long service in a hierarchical structure (Brezhnev in the Communist Party of the USSR, committee chairpeople in the U.S. Congress

4. Exemplary (role-modeling) -- leading by acting out desired behaviors (Gandhi, Thoreau, Schweitzer). It should be noted that some exemplary leaders have no immediately identifiable group of followers, but attract people across a spectrum of time or distance.

5. Communitarian (similar to exemplary) -- leading by working with a group of people to shape outcomes (Lech Walesa, while leading the labor movement, Solidarity; when he became President of Poland, Walesa's mode of leadership changed).

6. Technocratic -- leading through unusual insight into the power of technological change together with a mastery of some aspects of technology (Henry Ford -- in the early days of the assembly line, Werner von Braun in Project Apollo, which resulted in the landings on the moon).

7. Mediating -- leading by gathering people together around the persuasive power of enlightened self-interest (Dag Hammarskold, de Klerk in South Africa, King Hussein of Jordan).

8. Visionary -- leading by rallying people to move toward an image of a "better future," which may or may not be achievable (Adam Smith, Karl Marx).

In appropriate circumstances perhaps each mode of leadership served a useful purpose, while even the most benign form of leadership can also go awry in some circumstances. We note that even authoritarian leadership was perhaps appropriate at times when chaos threatened -- as with the rule of Elizabeth I in England at a time when both domestic unrest and the naval power of Spain posed serious threats to stability in England. Many of the leadership examples cited above, illustrate that "good" leaders have been able to bring about positive results through a broad variety of modes of leadership.

Transcending earlier modes of leadership:

Humanity has given rise to many and varied forms of leadership. Nevertheless, no leader has emerged to reduce the level or threat of conflict in the nuclear age. At best, leadership modes employed since World War II have postponed confrontations and created a generalized assumption that if we postpone major conflict often enough and long enough, there will never be another catastrophic, global war. The efforts of the many who forestalled major wars are laudable, but not necessarily repeatable.

We approach the task of suggesting a new mode of leadership in the conviction that "peace is possible," and at the same time we recognize that peace is a poorly understood process [Chestnut, p. 89]. We also believe that humanity can accommodate the biosystems without which human life would again become (as at the end of the Middle Ages) "short, brutish, and nasty."

The tasks are both daunting and achievable, and we posit that accomplishing these tasks will require a mode of leadership different from those seen before but yet patterned after the best the past has given us.

Prospects for sustainable peace growing out of the modes of leadership of Jesus, Buddha, or Mohammed are less bright than these leaders would desire -- because many of their followers have absolutist views and tendencies. Absolutist tendencies may motivate religious followers to fight to the death for their faith or to fight to the death for the annihilation or conversion of other believers. These absolutist tendencies too often result in war, no longer a serious option for survival, if it ever was.

It appears that a new mode of leadership must:

1. Transcend earlier modes of leadership;
2. Not be identified exclusively with a single major religion to the exclusion of others.
3. Be identifiable through a simple set of criteria.

A New Paradigm:

Four criteria seem especially crucial for a new mode of leadership adapted to the challenges facing humanity at the beginning of the Third Millennium, A.D.:

Vision, Creativity, Consensus-building, Providing incentives for appropriate social behaviors

Vision

Vision is the paramount attribute of leadership for the future. Taking humanity where it has never been, i.e., building a sustainable society, will require vision. Building social stability will require allowing supporting biosystems to maintain their own stability at sustainable levels. Clearly, a weakening biosystem could change in ways which would not support social sustainability. Sustainable social activity and sustainability of the biorhythms of underlying biosystems cannot be separated.

To envision how society will acknowledge its dependence on underlying biosystems will require unusual vision. It will be equally challenging to envision how competing and often hostile social subgroups can live together over long periods of time.

Creativity

Merely envisioning a sustainable dependence of society on its supporting biosystems will not be sufficient to make it occur. It will be necessary to create social mechanisms which will function effectively in this arena.

These social mechanisms must also enable competing, often hostile groups to live together for a foreseeable future. This will almost certainly require that groups -- political, ethnic, religious -- renounce any desire to dominate other groups.

Humanity has evolved an ethic which says, "dominate nature; dominate others." The new ethic, and particularly its modus operandi, will be shaped around a principle of "accommodate others; accommodate our supporting biosystems (nature)." A corollary to the principle "accommodate our supporting biosystems" might be: If we do not accommodate the biosystems which support us, eventually those systems will be unable accommodate us. For the totality of humanity there is no rejoinder to the corollary, though it might be possible for a small subset of humanity to live with very weak supporting biosystems.

Consensus building
If we assume that powers of vision and creativity sketch a workable social future, the new mode of leadership will be challenged to build consensus among various groups, many of which will:

1. Want to dominate
2. Lack relevant mental, moral, and social paradigms for consensus
3. Lack motivation to change.

The new mode of leadership must build a consensus among people and groups -- consensus around wanting to reach the envisioned future. Both the envisioned goal and the path toward that goal will be elusive. The new mode of leadership will realize this and will persevere until the goal is achieved.

The alternative is an unsustainable society and chaos. Clearly, consensus building skills must advance beyond levels attained thus far.

Incentives for the Future
The new leadership paradigm will be challenged to provide incentives for positive sum behaviors and disincentives for zero-sum behaviors and even greater disincentives for negative-sum behaviors, such as war and overstressing biosystems which support human societies.

5. CRITIQUE

The sketch of a new leadership mode is offered in hope that it will elicit alternative suggestions or modifications to help create a leadership style suited for the social clashes of the future and for the seemingly unavoidable weakening and possible collapse of key supporting biosystems. We welcome critiques of this sketch of a future mode of leadership.

At the same time, we feel it is inappropriate to dismiss this effort as naive and incomplete. It may be both, but to do nothing in light of certain social collisions and increased fragility of supporting biosystems if even more naive and very likely defeatist as well.

6. CONCLUDING NOTE

Analogical comparison of the new mode of leadership with an automatic control system

A vision of a desirable state is analogous to an input signal to a control system.
Creativity in structuring workable social interactions is analogous to the design of a control system.

Consensus building is analogous to furnishing a control system with power to create an appropriate output in response to an input signal.

REFERENCES

Chestnut, H. (1984). Uses of Systems Techniques to Improve International Stability. In: *Supplemental Ways for Improving International Stability*, Published for the International Federation of Automatic Control, Pergamon Press, pp. 87-92.

Novikava, S., Gancharova, S., Ivanov, A. et al. (1997). Hierarchical Mathematics and Cybernetics in Design of Sway Networks of States and States' Unions. In: Preprints of the International Conference on "Transition to Advanced Market Institutions and Economies". Warsaw, Poland, pp. 298-301.

POLITICAL, ECONOMIC AND MILITARY EVOLUTION OF INSTABILITY, NORTH KOREA, 1945-1998

Jacques G. Richardson

Decision + Communication, Cidex 400
91410 Authon la Plaine, France
e-mail: decicomm62@aol.com

Abstract: Korea since 1945 revives Le Châtelier's principle: if conditions in a system initially at equilibrium are changed, the instability will shift towards renewed equilibrium. Here we examine the nature of changing conditions in the years since the origin of the 'mental blockade' of the years 1953 until 1972, when the two Koreas began discreet, preliminary negotiations. Reviewed are North Korea's quest for nuclear energy and the efforts of other interested parties to limit the concomitant acquisition of sufficient plutonium that would facilitate the manufacture and deployment of nuclear arms. Borrowing an analytical template from Fisher and Brown, we trace the slow return of equilibrium within a political stand-off (Fisher and Brown, 1989). **[This paper should be read in parallel columns]** *Copyright © 1998 IFAC*

Keywords: distortion, crosstalk control, networks, feedback, decision-support systems, stability

1. AIM OF THIS FORMATTING EXERCISE[1]

To develop a rapport in which the adversaries can deal progressively with their differences. To do this, a clear perception is needed of the essential elements in their relationship. There are six:

- rationality (in an often emotional world)
- understanding
- systemic obstacles hindering communication
- reliability (or its lack)
- persuasion, and
- acceptance.

⇨ SEEN FROM THE NORTH ⇦

2. RATIONALITY (IN AN OFTEN EMOTIONAL WORLD)

N resents U.S.-Soviet partition of peninsula (1945) into ideological halves.

N seeks security, confidence, optimism; copes with hostility, rejection.

N insists on:

⇨ SEEN FROM THE SOUTH ⇦

⇨ Emotion tends to cloud judgment, ⇦ close doors to reason

ROK fears (after Second World War) replacement of Japan by USSR. Accepts U.S. as lesser evil.

ROK seeks security, confidence, optimism; copes with hostility, rejection.

ROK insists on:

---

[1] *Abbreviations: N, Democratic People's Republic of [North] Korea; ROK, Republic of [South] Korea; IAEA, International Atomic Energy Agency; ICRC, International Committee of the Red Cross*

- socialism,
- 'revolutionary struggle,'
- solidarity with international revolutionary forces,
- (later) *juche* or total self-reliance.

N: from 1945 to 1972, no Soviet-Chinese tutelage is too strong.

Beginning 1972, split N-ROK families may be able to reunite as a result of secret negotiations.

Answer to question: *possibly*

- free government,
- free economy,
- access to rest of world,
- membership in UN.

⇨ *Can emotions be balanced with reason?* ⇦

For ROK, from 1945 to 1972, no U.S.-"UN" tutelage can be too strong.

Beginning 1972, unofficial N links with ROK might help reunite families split by DMZ.

Answer to question: *possibly*

3. UNDERSTANDING

N: What is the outstanding element in thinking, action by ROK?

- ROK has made itself colony of U.S., dupe of UN, and not free of Japanese influence.

N: But our system must dominate that of the ROK.

N: Can we break into the world of ROK? Yes, by force, but we shall need support of USSR, China.

N: To be strong and remain so, we need to develop heavy industry, giving secondary role to agriculture.

N: During the 1960s and 1970s, any trade with ROK is out of the question.

N: Industry requires energy in the 1980s: USSR will provide petroleum, China coal, at preferential prices.

Perhaps the resolution of political and other problems will follow, but this is not a priority.

⇨ **We need to grasp contrast among systems** ⇦

ROK: What is the salient point in thinking, action by N?

- N remains Soviet satellite, conspirator with China, economically dependent of communist bloc.

ROK: But our system must dominate that of N.

ROK: Can we force ourselves on N? Perhaps not, but we need to maintain defenses against N's military strength.

ROK: To be strong and remain so, we need to industrialize and develop export markets.

ROK: During the 1960s and 1970s, any trade with N is excluded.

ROK: Industry requires energy in the 1980s; profits from exports will help our country buy energy abroad.

It is possible that resolution of all other problems will follow, but this is not a priority.

4. SYSTEM-INHERENT OBSTACLES

(N: Probable groundbreaking in 1979 of site for electronuclear reactor to improve energy supply.)

USSR insists (1985) that N sign Non-Proliferation Treaty in exchange for 4 Chernobyl-type light-water reactors. Prospects for this then diminish as USSR deteriorates (1988-1990).

Eduard Shevardnadze announces to N that USSR is ready to exchange diplomatic relations with ROK (1990). N threatens to speed development of nuclear arms. Shevardnadze counters that this would harm N's relations with West and international community (Oberdorfer, 1990a).

N is furious with Soviets/Russians and with the Chinese (1991-1992) over their new relations with ROK, then calms down.

⇨ **Obstacles that may impede communication** ⇦

(First detection in 1982 by U.S. satellite of this site, communication of discovery to ROK, UN.)

ROK, wishing to re-establish link with N, bides its time — despite multiple political assassinations by N — waiting for situation to improve during the late 1980s.

ROK President Roh Tae Woo asks President Mikhail Gorbachev privately (San Francisco, 1990) to tell N that Seoul is ready to meet officially to discuss any „outstanding" items. (Oberdorfer, 1997b). ROK establishes diplomatic relations with Russia, then China.

Despite all this „backing and filling" by both sides, ROK and N conclude a pact on 31 December 1992 in which the two sides decide to:

agree not to „test, manufacture, produce, receive, process, store, deploy or use nuclear weapons," nor to „possess nuclear-reprocessing and uranium - enrichment facilities." There is suddenly far less instability in this complex system of bilateral and even multilateral stress: consultation improves the relations between the two principal opponents.

5. RELIABILITY (TRUST)

⇨ **iteration begins. „Trust, but verify."** ⇦

(Trust between two parts of a country that has been subjected to some 1,800 incursions since about the year 1,000 B.C. has not been — between 1945 and 1998 — the most encouraging reality behind the development of a good working relationship between N and the ROK.)

IAEA asks that its inspectors have full access at Nyongbyon (1992), following U.S. detection of Iraq's secret preparations for atomic-biological-chemical warfare in relation to the Gulf war. N threatens to withdraw from NPT.

N permits, nevertheless, IAEA specialists to assure „continuity of safeguards" (May, August 1992). IAEA tells N (in November) of evidence it has of N's nuclear-waste storage at Nyongbyon.

ROK prepares (1992), with U.S., annual „Team Spirit" joint maneuvers. This continues years-long tradition of such exercises — in spite of ROK overture 20 years earlier to N for exchange of Red Cross amenities.

U.S. and ROK assess situation in N (May 1994) as one of constant military build-up, an „incremental normalism," yet one causing rising tension and watchfulness along DMZ, new and critical danger.

May-June 1994: Pentagon plans pre-emptive attack on N and the bombing of Nyongbyon. A plan of attack is elaborated. Option 1 - reinforcing ROK and U.S. forces in South Korea with some thousands more U.S. ground troops. Option 2 - adding squadrons of F-117 Stealth fighter-bombers and bombers nearby (probably Guam). Option 3 - sending a second aircraft-carrier and tens of thousands of army and marines to ROK.

N receives ex President Jimmy Carter across DMZ, at Pyongyang, with message from President Bill Clinton. Carter and Kim agree (15 June 1994) on nuclear freeze in N in exchange for 5-Mw reactor(s) and summit talks with ROK within month.

Relaxation of economic sanctions against N are announced by Washington (December 1994).

ROK profits from easing of tensions, although Kim Il Sung's death results in indefinite deferral of summit talks with N.

In March 1995, ROK joins U.S. and Japan to form Korean Peninsula Energy Development Organization (KEDO) to help N with energy substitutes to replace Soviet electronuclear plant.

6. PERSUASION

⇨ **The operating core of negotiation** ⇦

N begins (1993) using its nuclear effort as a „bargaining chip to trade for recognition, security assurances and economic benefits" from U.S. (Oberdorfer, 1997b). War is averted by Carter visit in June 94 (Sigal, 1998), a turning point in U.S. nuclear negotiations with N.

October 1955: N's GDP shows negative growth over years 1990-1994. As its military options shrink, *Time* magazine asks if N „will do something rash" (*Time* 1995). N continues missile development: No-Dong 1 should be capable of striking Japanese targets within coming year.

ROK makes various offers (1993-1994), some through China, to set up negotiations with N — without success.

Finally ROK signs accord to supply N with 150K tons of its rice free, but N prefers to hide source from its own recipients (1995).

ROK accuses N of reinforcing air units near DMZ.

ICRC says floods affect 5 million in N; 500,000 people are homeless (December 1995) throughout the north.

End of 1995: A young ROK military officer studying abroad develops thesis that a weak power, N, can successfully challenge strong nations *if* it weighs the situation sufficiently in its favor in terms of:

- **overall strategic calculations,**
- **its own offensive capabilities (small N still has more than a million me under arms),**
- **a strong-power alliance, and**
- **its domestic power-structure (Cheon, 1995).**

N receives (1996) first shipment of LWR components from KEDO. N becomes less sensitive because of a new „politics of hunger" (*Economist*, 1996). N examines possibility of U.S.-ROK-N-China, but China backs down.

ROK is mollified by U.S. as the latter grants new funds to World Food Program for aid to N (February 1996).

U.S. replaces its aging weapons assigned to ROK (April 1996).

7. ACCEPTANCE

⇨ Culmination of the iterative process ⇦

U.S. workers begin removing spent fuel rods from N's nuclear installation (May 1996). (See following page.) N accepts to help search for 8,100 missing U.S. troop remains in exchange for $2 million. N imposes severe rationing. N will receive $43.6 million in food aid via UN from ROK, Japan, U.S. and others (June 1996).

N's „salesmen" visit Japan to promote the Rajin-Sonbong economic zone in northern N (July 1996).

N suffers from more floods, mudlsides, crop damage (August 1996). With foreign exchange almost gone, N attracts Hong Kong financing for the Rajin economic zone (September). General stagnation of N continues, however.

U.S. permits Cargill to ship 0.5M tons of grain to N. N is now willing to discuss „framework for regional peace talks" (January 1997). N begins to accept nuclear waste from Taiwan (February) to earn foreign exchange.

Taiwan official visits N (July 1997), saying N shows no signs of attacking (*Newsweek*, 1997). N loses its ambassador to Egypt through defection to U.S. (August).

ROK's „meltdown" of its own economy leaves N „more calm and free of vitriol" than in many months (*Washington Post*, 1998).

N diplomat at UN's Food and Agriculture Organization defects in Rome to ROK. U.S. offers 200K tons of food to World Food Program for N.

U.S., Japan may have to foot bill for ROK failure.

N is quiet in early 1998, perhaps realizing at last that its instability may be finding new norms of equilibrium.

ROK approves nearly $20 million for three joint ventures with N (May 1996). ROK names its former Unification minister to head government party. ROK receives defecting N pilot, first since 1983. ROK officials see, however, little „popular unrest" in N (June 1996).

ROK professor admits espionage for N in ROK, after years of earlier service as Kim Il Sung's Arabic interpreter (July 1996).

ROK police quell „outlawed, pro-North" student rally in Seoul (August), but ROK proposes greater economic and cultural cooperation with N. ROK is invaded in September by submarine-borne infiltrators from N (who all die). N apologizes by December 1996.

ROK Embassy in China accepts defection by N's party ideologue, Hwang Jang Yop, and aide. ROK claims Hwang defected because of disgust with a socialism that has degenerated into feudalism (February 1997).

ROK ships 50,000 tonnes of food to N (August); ICRC will distribute. ROK, Japan, U.S. turn over $5 billion worth of power stations to N (August).

ROK has new president (January 1998), former dissident Kim Dae Jang, who asks N to join in „proper dialogue" with ROK and its allies.

ROK says it cannot meet its KEDO bill for reactors to N because of economic troubles at home.

ROK's new president may succeed with his *Nordpolitik*: „reach out to starving North Koreans and create conditions, if not for unification, then at least for peaceful accommodation" (*Economist*, 1998).

The statements illustrating the Acceptance phase of this model continue, as 1998 progresses, to describe the situation prevailing on the Korean peninsula after a half-century of acrimonious hostility between two nation-states of the same ethnic strain.
Le Châtelier's „law" may have proved to be inadequate in the realm of physical chemistry, yet it seems to have application in the type of large-scale systemic instabilities described here.

REFERENCES

Cheon, Jaeho (1995), an exchange student at the Naval Postgraduate School, Monterey CA, in discussion with author concerning his unpublished dissertation.

See also:

Courrier International (1996). French weekly publishes cover story on „North Korea: Last Stalinist Hell." 17 October, pp. 5-8.

IAEA director-general (1997) reports „canning" of

Economist, The (1996), 27 January, p. 54.

Economist, The (1998), 31 January, p. 63.

Fisher, Roger, and Scott Brown (1989) of the Harvard Negotiation Project, *Getting Together, Building A Relationship that Gets to Yes*. London, Business Books Ltd.

Newsweek (Atlantic ed.)(1997), 28 July, p. 30.

Oberdorfer, Don (1997a), *The Two Koreas, A Contemporary History*, Reading MA, Addison-Wesley, p. 213.

Oberdorfer, Don (1997b), *op. cit.*, p. 212.

Oberdorfer, Don (1997c), *op. cit.*, p. 305.

Sigal, Leon V. (1998), „Jimmy Carter Makes a Deal," *Bull. Atomic Scientists*, January-February, pp. 40-46; *cf.* Sigal (1998), *Disarming Strangers: Nuclear Diplomacy with North Korea*, Princeton NJ, Princeton University Press.

Time (Atlantic ed.)(1995), 20 November, p. 27.

Washington Post, The (1998), reprinted in the *International Herald Tribune*, 7 January, p. 4.

irradiated fuels at Nyongbyon almost completed. Media release PR 97/32, 14 November.

Le Monde (1996). Own dispatch claims Kim Jong Il bides time two years after father's death because he wishes combined industrial, energy and food crises to blow over, 7 July, p. 4.

Oda, Ko (1997) of *Asahi Shimbun* reports that 12 members of Kim Jong Il's extended family occupy positions of authority in N. Cited in *Courrier International*, 3 April.

Ogawa, Sinichi (1997). Researcher at National Institute of Defense (Tokyo) proposes Nuclear Weapon-Free Zone in Northeast Asia. *International Herald Tribune*, 29 August, p. 8.

Washington Post, The (1996). In interview former U.S. Ambassador to ROK James R. Lilley advises his countrymen not to „waste money on North Korea's rapacious regime," reprinted in *International Herald Tribune*, 22 July, p. 6.

Wilson, Jim. *Popular Mechanics* (1997), quotes Douglas Bandow of Cato Institute: U.S. should leave ROK because latter has national income 18 times that of North Korea. March, p. 47.

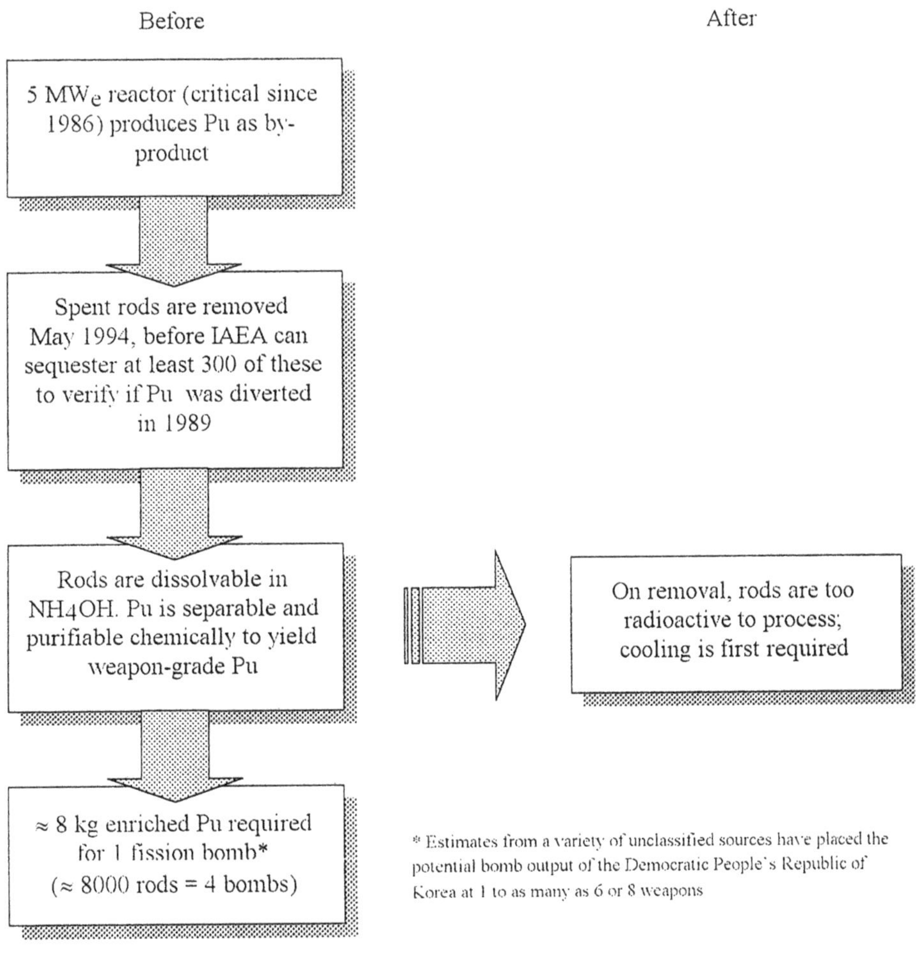

Fig. 1 Functional decomposition at Nyongbyon (1980s-1990s)
Source: Richardson, J. G., „Decision Processes and the Acquisition of Nuclear Energy by the Democratic People's Republic of [North] Korea," Proc. 1995 IEEE International Conference on Systems, Man and Cybernetics (Vancouver).

VIRTUAL ENTERPRISE - A STRATEGICAL APPROACH TO IMPROVE SOCIO-ECONOMIC STABILITY

I.Dumitrache, A.M.Stanescu, S.I.Caramihai, A.Curaj

POLITEHNICA University of Bucharest, Human Resources Training Centre
132 Calea Grivitei, 78122, Bucharest, Romania

Abstract: The new manufacturing paradigms, starting with agile manufacture and lean production is loading our civilization towards the new generation of concurrent enterprises involving virtual enterprise architecture and concurrent engineering extended methodology. The present paper aims to identify the impact of these new concepts on the socio-economical life and to underline the role of open universities in the formation of virtual enterprise-oriented specialists. *Copyright © 1998 IFAC*

Keywords: virtual enterprise, fast-reactive, business process reengineering, open University

1. INTRODUCTION

"The world in which we live may be divided into the natural world and man-made world" [Prasad '95]. Since XVth century manufacturing has provided this man-made world with all products made by people for the use of the people.

Manufacturing has seen many times evolution through revolution. The latest iteration was the Information Revolution. The manufacturing managers are besieged with new ways to create products and survive in the market place.

The core of this revolution is the concept of making information available to all who need it. The aim is to empower the humans at all levels of the company, allowing them to take vital decisions on the improvement of quality, productivity and responsiveness. [Beakly, 86]

Many companies use a narrow definition of manufacturing that undermines their efforts to be world-class, i.e. „manufacturing is the production activity on the shop floor". But, in order to be a world-class performer it is necessary to consider manufacturing in a more holistic manner and to include the complete spectrum of activities that starts with product and process design, continues with production planning and control, flexible production, distribution a.s.o. for ending with after sale service and support. [Blanchard, 76]

There has been a renewed interest in improving manufacturing during the last quarter of the 20th century as nations have re-discovered the truth „*no wealth is gained unless something is made*". The development and production of industrial products is currently often released by a **network of cooperating enterprises**, together forming dynamic extended organization within market framework. One could view this new type of distributed enterprises as a business integrator, having a specific life-cycle of business duration. Each node has to perform a dedicated task, based on their own specialization in the total business process [Hof, 97]. By concentrating on its care-competence each individual enterprise will depend strongly on the effectiveness of functioning within such a network.

A challenging domain for concurrent organization science is now-a-days focusing general attention of both scientists and decision makers.

Despite of various acronyms, circulating all over the scientific world (DDE – Dynamic Extended Enterprise, NE – Networked Enterprise, DE – Distributed Enterprise, VE – Virtual Enterprise) a solid sound trend could be identified. It is concerned with a new Information and Communication Technology (ICT) – based organizations, acting on Global Market within Information Society. This entity is faster reactive for business process triggering than well-known traditional organizations (large enterprise or SMEs).

In our opinion, Virtual Enterprise is the most appropriate denomination for the new concept due to its basic meta-system features:

1. **business integrator**, allowing both the vertical integration of SMEs [Santoro 97]
2. cybernetic communication space **user** for electronic commerce, business process
3. multi-agent based internal **coordinator**, implementing Production Planning and Control (PP&C) within wide-area-network of heterogeneous platforms
4. concurrent engineering oriented **designer** for product/ process facilities according to virtual prototyping philosophy

The present paper is aiming at focusing the attention of decision-makers on the potential force of virtual enterprise, acting as stabilizing factors on the Global Market.

2. VIRTUAL SPACE BASED NEW „WELTANSCHAUNG"

Virtual space is a new Information Technology based concept which refers to an age-old meaning. However, the revolutionary character of this new term finds its force only in its original context.
The human being has always had two spaces in which he could „move":

- the objective extraneous world
- the space issued from his/her own projections/ representations

A projective world, underlying the objective world has been created since the very origin; it was at the beginning a feature of *myth* and the *sacred* in general, but it was used also as *a way* of establishing numbers by Pythagoras brotherhood, or as the way of writing Newton's theory, though as a scientific approach to understand and study the real universe.

Virtual projections of the real world are used not only for the study of its particular phenomenon and/or rules, but also for modeling social behaviors. The most important exemplification is the school: school evolves in a virtual reality were disciples are „classified" not only according to their native intelligence, but mainly according to their capacity to adapt themselves to behavior rules and informational models used in a given school.

The increased complexity of models used in virtual world imposed the creation of appropriate tools. One of these tools was the computer, that evolved gradually, becoming a medium of interaction of the projective subject not only with a virtual world, but also with him/herself and his/her fellow beings.

This tool allows to create the IT-based virtual world which represents a connection bridge between different people and even communities as long as this virtual world is able to exist as a creative and reality establishing space.

This projective world as a substitute of the objective world ranks together with the world of symbols governed by Mathematics under the focusing force of technologies, creating a multi-dimension universe which is both rigorous and polyvalent, in which the individual may find him/herself not as a lonely demiurge but as an essential partner in the Creation play.

Within the context of modern virtual universe, the passing from the scholar's creative singularity to the synergism of the partner in a collective creation represents the main task in reshaping the educational approach and contemporary research and production methodology.

In this space, the virtual enterprise appears as a particular but major concept. It can be generalized as representing any possible „enterprise" and can be viewed as the very model of creation, namely of the way in which one can actually participate in the virtual space.

Once the initial special status is removed, together with the precious and mysterious halo, these concepts regain their value and force which open the path to a new era. This opening requires not only the elaboration of a new work method in the virtual space, but much more than this, a new educational universe - formative and educational - to enable the individual have access to the new status offered which, however, tends to escape him even when he/she is a practitioner in the new, virtual field.

Thus, before establishing new trade and production relations *in* and *through the medium of virtual reality* , we have to build a new Weltanschaung, a new motivational universe, a new self- and world-image of the individual, new conceptual valencies, new dialogue skills. Paradoxically, the border between learning and practice, between play and business is not going to become more prominent but be rather blurred.

The individual's break with isolation and the transformation of the dialogue, if we don't let ourselves be cheated by the apparent ease, represent a leap whose extent, problems and traps are still quite difficult to perceive.

Virtual Enterprise becomes a reality which could no more be ignored as impacts on our civilization, especially because it was imposed by practical socio-economical necessities. This reality could improve life conditions in the whole world, if good managed. but could also present some risks, if only parts of the concepts are considered.

3. VIRTUAL ENTERPRISE - THE BASIC CONCEPT

The materialization of the paradigm of virtual enterprise (VE) although enabled by recent developments in communication technologies and computer networks, requires the definition of a reference architecture and the development of a supporting platform and appropriate protocols and mechanisms.

There is no unique definition of VE, but those usually used permit the underlining of its basic characteristics, as follows:

Definition 1: A Virtual Enterprise is a temporary alliance of enterprises that came together to share skills and resources in order to attend a business opportunity and whose cooperation is supported by computer networks and adequate IT tools and protocols.[Afsarmanesh, 97]

Definition 2: The Virtual Enterprise is a temporary alliance of partners focalized on their respective core competence that is capable to assure the cooperation as a support for the network innovative process and for a rapid response to business-market requirements.

Definition 3: The Virtual Enterprise could be defined as an open meta-system in a business universe where acts a set of fabricators utilizing multi-model representations of products/processes and cooperating in the framework of temporary alliances for achieving a set of goals:

$$VE = (U, F, Mv, G)$$

where:

- U is an universe of discrete business processes
  Note: The CIM-OSA methodology was the first to formalize business as a specific type of processes triggered by market requirements.
- F is the set of fabricators
  Note: A **fabricator** is an autonomous agent having a set of core competence *i.e.* a set of functions that it is able to accomplish. A fabricator could cooperate with other fabricators with similar/ different competence by communicating goals and data. By modifying its own performance evaluation measures according to a global goal and optimality criterion, a fabricator defines itself as a fast-reactive component of a meta-system.
- Mv – is the set of representations of products/ processes/ facilities depending on the type of fabricators utilizing them *i.e.* the set of multi-model representations [Stanescu, 97]
- G is a structure of goal-type that is decomposable into a tree of objectives for every fabricator implied in the VE meta-structure

As definitions underline, the main features of a VE are:

- the capacity of innovation
- the capacity of cooperation
- the capacity of internal reconfiguration
- fast-reactivity
- open character

The interconnection of partners-fabricators raise the following category of problems:

- communication (infrastructure, information security, a.s.o.)
- interaction management (based on the standardization of the information exchange)
- selection of the partners and **configuration/ reconfiguration** of the alliance

The configuration of a VE is the result of a negotiation concerning a particular competition. A **competition** is a triplet composed by:

1. a global goal
2. a metrics for evaluation of the results/ competitors
3. a set of restrictions in the achievement of the global goal

The configuration by negotiation of a VE has two main phases:

- *The preliminary configuration of the alliance*: for the participation at the negotiation as a bidder, when the alliance itself is virtual and its global goal is to win out the competition; it participates as a **business integrator**. When the competition is finished, the winner became the global business integrator and its global goal changes into the global goal of the VE

- *The configuration of the VE operational alliance*: the aim of this phase is to refine the VE global goal structure into objectives for every fabricator implied so as the alliance to succeed in offering the required services at required moments and within the imposed cost limits. The **global business integrator** decides on the necessity of reconfiguring the alliance, according to the performance analysis of the VE.

A possible reference architecture for such a virtual organization is represented in fig.1.

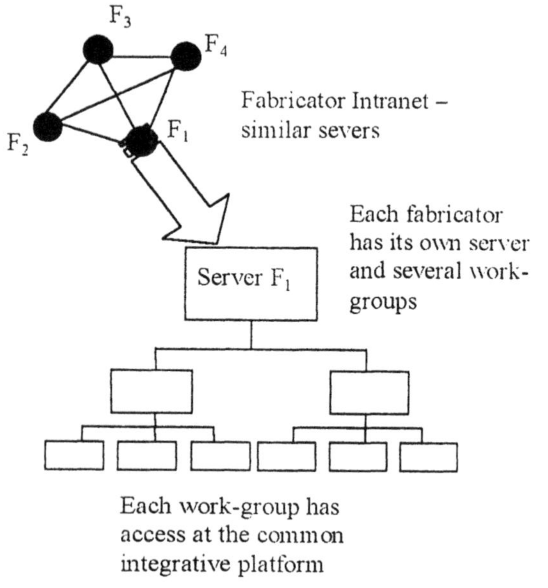

Fig.1 Reference architecture for virtual organization

4. THE FORMATIVE IMPACT: EDUCATION CHANGE MANAGEMENT

The VE represents a new kind of work-place and people operating in its framework must provide certain qualities both professional and social. The appropriated formative place for VE team-specialists is a dedicated open-university.

The VE oriented open university should be viewed as an interactive, modular, integrative, adaptive network program. It must allow the disciple to gradually integrate new knowledge, to formulate new problems and bring into new focus each of the detail issues tackled earlier.

They should submit these problems to their network colleagues and respond to those submitted by these colleagues. The students from the lower grade should get used to the existing models and with the logic of this type of approach, those from the upper grade should already integrate new knowledge and approaches carrying out new models.

Such a program then appears as educationally formative, both when used and/or written. The interaction educative character affects those who are learning to use it, those who are learning to write it, as well as those who are trained as moderators, as project managers, teachers respectively.

An oversimplified example of study modules is representing the analogy with the design phases of an electric circuit:

- in a first module, the program provides the electric components set: resistances, diodes, gauges, sources, etc. The student can form various circuits with them by coupling individual graphic icons. A second module enables the student to select the equations which are necessary for the circuit analysis and to study its characteristic curves. A third module will enable the student to actually study the connection difficulties, with additional contact resistances, etc.

Then the modules are diversified on various directions:

- making a compact assembly within an apparatus or device
- optimization of the ergonomic aspects of the control panel, of the assembly technology, of the aspects concerning reliability-costs ratio
- cost estimation
- study behavior in exploiting, reliability, errors
- study the local behavior of the various components
- programmers for error statistics, microscopic behavior of materials a.s.o.

Each module is open, permitting a free coupling to any program package. Thus, a self-evolving and self-improving laboratory can be created and gradually improved to the effect that each student working with these modules raises new questions to be solved by the students in the upper grade. At the same time, this continuous activity is first meant to be formative, to discover the programs' limitations and errors. During this training, in the second stage, certain functions of the programs are systematically altered and the students have to discover and correct them. However, this activity also helps correct the involuntary errors made by programs makers and which were not detected by the coordinators.

Such structure of education would give a new motivation to the student: it will prepare him for teamwork, it will provide him with initiative. communication skills and it will make him used to the virtual universe.

Such a system is self-sustaining as it represents a real educational software 'factory' in and for the virtual universe as the student is learning and checking these programs. imagines tasks which he will solve in the upper grade.

The modular character permits the successive tackling of problems as they are assimilated and their immediate integration, which is specific to the concurrent engineering approach. Thus, their concurrent engineering training is virtually synchronous with their basic professional training. The open modular character also helps to create a virtual laboratory in which the courses, seminar and laboratory activity tend to merge. Thus, it should be emphasized that in this inclusive model the basic notions of mathematics, physics, chemistry as well as those of marketing, economy and management should also be modularly structured according to open programs which may be interconnected. If we also add the specific design issue, work and consumer psychology, social issues a.s.o., we notice that the university might become the laboratory in which through the very educational act. with the same intellectual efforts and funds. tomorrow's virtual universe is being actually created.

The student is to be trained in the virtual laboratory which presupposes from the formative point-of-view. the virtual factory.

It is important that, beside the initial difficulties. if we consider education as a whole, from the lower grade to the post-graduate and to doctorate levels. global costs are even decreasing with efficiency is tremendously growing.

It is true that this is an initially slow growth and that it entails the risk of the inherent mistakes of all beginnings. Only after a certain period the results become significant and the growth obvious.

5. FACTORS IMPROVING GLOBAL STABILITY

Beyond the formal multi-modeling based on both Discrete Event Systems (DES) and Intelligent Agents (IA) techniques there are some significant socio-economical factors within VE framework. improving global stability.

S1: **Empowering human resources**: VE-oriented Open University

According to the change management of the global market pressure factors pushing on traditional university the new coming educational system is self-sustaining but it requires clear determination in launching the educational revolution and an international supporting and coordination effort. The main objective is aiming to provoke- new menthal approach for introducing market-like model and competition-oriented motivation for all categories of students. They will be able to get knowledge and skills based on team-work approach for educational software factory. They will develop collaborative work within cyberspace. The expected consequences will involve a smooth integration in VE-oriented engineering paradigm.

S2: **Improving worklife**: VE based creative engineering

A trivial solution could involve developing of home-jobs. but more important is to promote a new generation of micro-enterprises integrating few (3-5) "best in class" employees. acting during certain phase of virtual product or system development. By general dissemination of business information, a new job-market including teleworking will create as many as necessary jobs in terms of well-balanced regional economy. aiming at an efficient resource management.

An useful cross-cultural civilization oriented towards capitalizing and capabilities improving should be developed within multi-national VE framework.

E1. **Increasing SMEs stability**: VE-oriented organizational structure

New global market opportunities and innovative product development require financial and technical capabilities that individual SMEs can hardly provide [Santoro. 97]. Japanese manufacturers developed an integrated subsystem for earth moving machines, using components supplied separately by different SMEs and assembled directly by the final client. This resulted in the elimination of the SMEs from that product market. Large companies are in fact seeking for delegating development and production of entire subsystems to reliable suppliers sharing the business risk.

A short list of VE requirements should include:

- adaptivity facing the global market. based on fast reactive organizations
- internal flexibility based on dynamic reconfiguration capability
- global efficiency based on resources management

- dynamic controllability based on business events monitoring and supervising

E2. VE-based dynamic coo-petition enterprising (cooperation / competition)

The basic competitiveness is going to become the best-in-class selection process such as the intrinsic economic law of free market will promote the most reactive VE with respect to market changes.

Each partner within VE network of excellence will capitalize the field knowledge according to their core competence identifier. The global stability will be determined by stabilizing sub-systems that could be tested along a business life-cycle.

CONCLUSION

The paper is a first attempt to cross the border of specific scientific approach for virtual enterprise basic concepts.

The new manufacturing paradigms, starting with agile manufacture and lean production is loading our civilization towards the new generation of concurrent enterprises involving virtual enterprise architecture and concurrent engineering extended methodology.

Goal fixing for XXI century should involve not sustainable economical organization only but empowered socio-economical organizations to improve global stability within information society.

REFERENCES

Afsarmanesh H., Garita C., Hertzberger L.O., Santos-Silva V. (1997). Management of distributed information in Virtual Enterprises – The PRODNET approach. *Proc.of the 4th International Conference on Concurrent Enterprising*, Nottingham UK, pag.231-240

Blanchard B.S. (1976). Engineering Organization and Management. *Englewood Cliffs, New Jersey, Prentice Hall*

Beakly G.C., Evans D.L. (1986). Engineering: an introduction to creative profession. *New York, MacMillan*

Hof K, Pallot M. (1997) Fast Reactive Extended Enterprise. *Proc.of the 4th International Conference on Concurrent Enterprising*. Nottingham UK, pag.339-345

Prasad B. (1996). Concurrent Engineering Fundamentals. *New Jersey, Prentice Hall*

Santoro R.(1997) The Virtual Vertical Enterprise approach for SMEs. *Proc.of the 4th International Conference on Concurrent Enterprising*. Nottingham UK. pag.361-363

Stanescu A.M., I.Dumitrache, A.Curaj (1997) Multi-modelling based system design within virtual factory synergetic framework. *Proc.of the 4th International Conference on Concurrent Enterprising*. Nottingham UK, pag.173-183

SOCIAL ENTROPY, TECHNOLOGICAL CHANGE AND INTERNATIONAL STABILITY

Smaranda Dobrescu*
Radu Dobrescu**

*Parliament of Romania, Chamber of Deputy
**POLITEHNICA" University of Bucharest, Faculty of Control and Computers
313 Splaiul Independentei, 77206 Bucharest, Romania,
E-mail: radud@dracula.aii.pub.ro*

Abstract: The paper presents an entropy-based model for a self-organizing hierarchical social-economic system. Different formalisms are analyzed, and the most suitable, based on a complex structure with decomposable levels having elementary actors as resource transfer units was proposed. The crucial observation is that a self-organizing system will tend towards differentiation along dimensions which are functional for its survival and consequently complex dynamic systems can be thus compared in terms of their evolution. Particular entropic concepts are adapted in order to models the dynamics of an economic-social system. Based on this model, several complex relations are presented, pointing on the importance of the technological change (and especially on the academic-industrial-economic-political relations) in the progress of the society. Finally, a possible global (world view) model with high stability presented as a generalization of several local entropic subsystems is discussed. Two typical conflicts with international consequences and the way to solve them are used as examples for the opportunities offered by the global model. The concluding remarks offer six basic implications which allow to develop further researches.

Keywords: social entropy, hierarchical systems, evolutionary economics, resource transfer units, stability, world view

1. INTRODUCTION

In the last years, many laws of physics are applied to the analysis of social-economics problems; their acceptance is at risk only when some new fact turns up that is not explainable by the recognized theory.

Among these laws, the Second Law of Thermodynamics (the Law of Entropy) has presented an extraordinary interest along the years and is largely accepted in various areas of knowledge such as Psychology, Sociology, Communication Theory etc. The last of this named domains is very important due to the informational approach of entropy, proposed by Shannon, which offer a general perspective on the connection between entropy and the structural order of a system.

The extensive application of the laws and methods of physics requires some caution as human behavior is ruled by more restrictive criteria of ethical, social and religious character. However, there are special situations in which is possible to draw generic conclusions about phenomena that involve humans, in general those situations in which physical conditionings are very strong, such as some problems of production and consumption. Since production involves among other things energy, the laws that govern the energy correlation also describe aspects of the production process. From this point of view, the economic problems seem to be strong connected with the energy and consequently with the thermodynamically entropy, while the social problems appear more connected with the aspects of stability, order and hierarchy proper to the informational entropy.

We propose here an evaluation of economic trends from the point of view of entropy and some speculations about the effects on organization of society, especially due to the technological changes, having in common their influences on the local (national) and global (international) stability.

2. A GENERAL ECONOMIC-SOCIAL ENTROPIC MODEL

The answer to the question if contributions of economics, sociology, evolution and system theory can be relevant to the society dynamics depends of the relevance of a complex model which enables to proceed from theoretical specification to generalization. "Variation" and "selection" should then be considered as sub-dynamics of this complex model. By understanding variables as fluxes, is necessary to distinguish change in the value of the variables from interactive effects affecting the variables themselves. Several approaches are possible: first, is possible to adopt a system with feedback, which implies a coevolution between the macro-system and the micro-variations closing the evolutionary system into a self-referential loop; second, an algebraic model allowing an algorithmic simulation can offer an additional degree in freedom of variables characterization, since variables can be considered as instantiations of fluxes; third, a multivariable model with nonlinear dynamics can be considered as another partial perspective. In fact, in this paper we propose a general model which offer the possibility to compare the various combinational solutions in a self-organized system using an entropic approach. The basis of this hypothesis resides in a "classical" assertion (GEORGESCU-ROEGEN, 1971) that "all entropical system self-organize their thermal death", i.e. the potential energy contained in the initial conditions dissipates in the system as rapidly as possible according to the dynamics of the system. Such a system perform life-cycles in a hyperspace that include the time dimension. Differentiation is then the equivalent to decomposability among the time axis. Self-organizing systems will tend towards differentiation along dimensions which are functional for the survival of the system. Complex dynamic systems can be thus compared in terms of their evolution: the more differentiated they are (that means a minimum of the global entropy), the more complexity the system is able to hold stable. If the global (higher-order) system goes into crisis, the subsystems becomes less controlled and thus more active.

So, the dependencies and interdependencies of the local processes can be modeled to describe how micro-events evolve into hierarchically ordered, macroscopic structures of time-conditional probability distributions. For example, if we let the configuration of structural units in state space be the N-dimensional vector x at time t, given the probability density function P(x,t), we can derive a macroscopic equation that describes the evolution of the transition matrix as:

$$d\langle x\rangle/dt = f_l[\langle x\rangle;P] \qquad (1)$$

where f_l is a nonlinear polynomial.

The apparently uncoordinated interaction of microscopic elements leads to emergent macroscopic properties that can be represented by a small number of order parameters that summarize the net result of complex system feedback and dependencies. Higher-order operations tend to reduce the transformative complexity of the sustaining lower-level processes such that, from a starting set of initial conditions of dimensionality N, the system locks in a much more compact state with lower dimensionality.

Time-dimensional configurations of such transformative processes are essential features of evolutionary theory in all areas of inquiry. Collective properties of these transformative processes should be describable as moments of a probability density function that substitutes for the microscopic dynamics. Consequently, a hierarchical level is determined by the number, location, complexity and degree of operational interdependence among components that maintain flexibly bounded, ordered relationships among themselves. Probabilities that any specific macro-ordering of relationships between individual components will obtain can be calculated using a generalizable formula for entropy (such as the Kolmogorov - Sinai equation; in fact we prefer between different measures of "thermodynamic" versus "informational" entropy a more generalisable interpretation of entropy as a measure of the probability of obtaining any specific macro-state given a particular number of micro-states). In this aim the relationships between components can be expressed as transitional probabilities in a tree diagram, as Markov and semi-Markov chains or as a coupled map lattice that models the distribution of states.

For this proposed model, the simplest way to describe the dynamics is to define a basic process of evolutionary change; we consider in both economical an social systems this basic process as being the *resource transfer*. Considering resources anything that interactants use to sustain interaction, the basic unit of change is the "resource transfer unit" (RTU). The resource interactants that constitute RTUs may be individual or collectives. The RTU is conceptualized as the smallest autonomous component of system operation that could serve as the basis of self-organized evolution at all levels. Types of resources, means of transfer, rates of transfer and transfer redundancy are selected so that an RTU can be defined as a relationship between at least to interacting entities such that over time the same resource is repeatedly transferred at the same rate by the same means.

RTUs concatenate to form chains or lattices of resource dependency configuration that define a basic transition matrix. Interrelationships between RTUs comprise self-organizing hierarchies. As the system transform resources, some transformations are inherently asymmetrical and cannot revert to an earlier form. New interactions appear, while resources combine and recombine, so that products of resource interaction are added back in the system as augmented resources. Resource augmentation in turn affects the augmentation development and depletion of resources throughout the system. Since resources are interconnected in entire chains of resource transfers, the interaction effects of the additions reverberate throughout the system. As augmentative interactions appear, resource configurations change and the system evolves. As a system evolves, there are thresholds of change beyond which entirely different resource conditions and resource configuration emerge; so, adaptation is a process of continual self-selection.

So, in complex hierarchies entailment structures form when resources are transferred between interacting entities. Since the transfer of resources among one transfer unit is related to the resource transfers of other units through common connectors, some groupings will form subsets of other groupings. In order of complexity, we can use the hierarchy with five levels proposed by FREESE (1988): from units to groups, from groups to unions, from unions to associations and from associations to assemblies.

With the resource array and the components of intermediate transfer hierarchies, the necessary and sufficient conditions to maintain stable a system of self-organization are the ability of this system to simulate an environment as well as to simulate parts of itself. However, in order to limit behavior and therefore the number of possible system states, a coupling between levels via specific resources transfer and feedbacks is necessary. The coupling between any two levels can be described by a joint probability function that makes explicit the mutual information (transinformation) contained in each level. This mutual information is expressed by the following relation:

$$I(t,m;i) = \int \log p(x_n(i)dx_n(i) + \int \log p(x_{n+t}(i+m)dx_{n+t}(i+m) - \int \log p(x_n(i), x_{n+t}(i+m))dx_{n+t}(i+m) \qquad (2)$$

where $p(x_n(i),x_{n+1}(i+m))$ is the two-point probability function and $p(x(i))$ is the single point probability function.

The behavior at each level will be limited by the actual interactions among RTUs as well as by the dynamics of the higher order grouping. At each hierarchical level, the states of levels above and below act as bounding conditions for the behavior of variables in the level under consideration. Each resource set in the base array, along with its concomitant intermediate hierarchies, would be defined as a *decomposable level* (DL) of the system structure. Additionally, each level within any given intermediate hierarchy would be an NDL. The problem of modeling macroscopic evolution thus becomes a description of the coevolution of two or more levels.

To develop such a model, one would need to be able to specify the functions that describe the relationship between rates of RTUs behavior, i.e. activation/degeneration, resource augmentation or resource depletion. These rates would differ for each DL, with lower-order levels behaving asinchronously in relation to higher-order levels. Such a function might be similar to the logistic equation (KANEKO, 1991) which simulates a globally coupled system of interacting nonlinear elements in a "hyperchaotic" systemic state.

In fact, our entropic approach is compatible with models for the hierarchical dynamics exhibited by differing dimensionalities in mathematical chaos (MORRISON, 1991). For example, when RTUs chose between two or more equipossible resources, as a result appears a "decoupling" of the association level; that decoupling will produce a feedback effect to lower levels and we assume that, although the transformation processes are stochastic, the feedback effects are deterministic. These will be especially true as we look to the hierarchy of the social levels of the system. At the level of society, human resources transfers can effectively speed the rates of resource augmentation. The aim of this paper is to establish, using this model, a chain between primarily resources, technological resources and human resources in our particular society, often named society of transition.

3. RELEVANT RELATIONSHIPS BETWEEN RTUs

For this period of transition, the term of sociology of translation (CALLON, 1992) seems very appropriate. The starting point in this theory is the importance of the technology and of the scientific knowledge in the non-equilibria dynamics of evolutionary economics. The sociology of translation considers two mechanisms: one for association (the primary mechanism), other for attribution (the secondary mechanism) as emerging from action governed by uncertainty. In term of empirical research, the foregoing is elaborated as follows: first, a problem is taken as the starting point (the actor's definition); second, all elements of the problem need to be tested and aligned (the translation procedure); third, each element receive an action to do (the enrollment process) and fourth, association for renewed actions are performed (the mobility aspect). The unit of analysis of the sociology of translation is, in conclusion, a discourse (i.e. a communication network), whose construction is based on entropical laws.

3.1 *Basic relations between economy, industry and energy.*

The classical economic models do not include energy explicitly among the production factors even though they make some reference to energy parameters. Nevertheless, there is a practically linear correlation between the product of an industrialized country and the use of energy. Even in preindustrial economies there is some correlation, as shown by the coincidence of economic booms and the introduction of new sources of energy. Presently we watch the world effort to keep oil prices at levels compatible with economic international stability. Mainly in countries of small revenue, the control exercised by governments over energy prices is a tacit recognition of its strategic importance. Another important relation appears between the productive system and the environment in which it operates. While production occurs in steps, wastes do not come out from one single point. The importance of the problem of availability of energy results from the fact that 90% of the energy used in production comes from transformations that go to heat production. Even tough the energy stock remains constant, the availability of energy decreases, due to the entropic irreversibility which occurs in the interaction of the productive system with the environment.

Hence, if the system whose transformation liberated energy (mechanical, chemical or nuclear) passed from a state generically called state U_1 to a state U_2, the work to which this energy is converted is:

$$W = (U_1 - U_2) + Q - I \qquad (3)$$

where Q is the quantity of heat exchanged by the system with the environment and I is the effect of irreversibility. Q is positive if the system receives heat and negative if it gives off heat. I is always positive, what means that every energy transformation signifies loss of availability by irreversibility (external, internal or both).

A similar equation may be written to describe the transformation of raw material in the production line, changing the sign of W and of $(U_1 - U_2)$ but keeping the sign of I. Therefore, the loss of availability occurs in the production of work as well as in its use in production. We know by experience that two motors that receive the same fuel can produce different works, in the same way as two enterprises that operate with the same raw material, the same energy resources and the same technology may have different costs according to the age of equipment, the managerial practices, the stocks, the man power qualification, etc.

In order to apply the exposed considerations to analyze a productive system, it is convenient to consider $(U_1 - U_2)$ a product of the fuel mass used and the calorific power (for other forms of energy the names are different but the concepts are the same). So, we obtain:

$$W = m\,H + Q - I \qquad (4)$$

We should observe that the terms in equation (4) may be annually balanced as it is done with the National Energy Balance, so that they can represent work production rates, energy resources utilization rates etc.

If we admit that the population is constant and that the type of products consumed are not changing, we would have a situation called steady state (of production and consumption). But, like natural resources, energy and raw material are finite and, since the economy of production leads to consuming first the best resources (those of larger concentration of the material of interest and more accessible), to maintain the permanent regime is fictitious, since more and more the resources will be farther and of worse quality. So, the work required for production increases, what requires an increase of energy resources consumption or the use of resources with higher concentration, or a decrease of Q or a decrease of irreversibility.

3.2 *Relations between human factors*

The main difficulty in implementing these measures is the involvement of the human factor in the question. Specialization opposes the use of creativity, hierarchy makes difficult the relationship among people at different levels of responsibility, automation and informatics substitute man for very specific equipment, causing unemployment and deterioration of social relations. Even with a stable population and with consumption qualitatively unaltered, it is not possible to maintain a permanent regime in economy. The possibility of a crisis appears when economic forces are more and more tending to split society in two: between those who have good jobs and whose standards of living continue to rise and those who are faced with either falling incomes or the prospect of more-or-less permanent life on the dole. The non-rationalized economical activity provokes unbalances whose negative results are already being felt in the unemployment and in the worsening of social inequalities, with the forced reduction of the workers, consumption due to unemployment or to subcontracting. In the same way, globalization brings denationalization of the economy which may cause a lesser use of local potentialities.

Our entropic model, when applied to society, can give us concepts of the social conservation and social entropy law with its necessary dissipation of nonreturnable energy, social friction, and so social matter social energy equivalents. Within the concept of the quantification of the losses, in depth studies must lead to the determination of corresponding friction losses. Here we have to count on the unpredictability of human creativity and inventiveness to supply us with social lubrication to reduce the social friction.

Successful application of the above approach, will result in greater conservation of social energy and the slowing down of social entropy. For this we need to concentrate on some deeper grasp of our subjective-objective constructions and on the quantitative assessment of the social world. Very promising results in modeling social relations and analyzing the social stability are expressed in a row of papers of Iacobas, from which the most prominent (IACOBAS,1995) offer a complete thermodynamic approach of a social system, using the concept of sociomol to define the dimension of social groups involved in RTUs and the social temperature as a parameter of the social interaction. We use some suggestions in this aim for a better exploitation of the academic-industry relations.

3.3 Academic - industry relations as sociological solution for economic development

One of the most important RTU is proper to the technological change, as being the principal cause for well-thought-out plans for economic, social and institutional changes. A strong relationship was identified between the amount of resources committed to research in one area of technology and the rate of technological change. But the main question is how technological change takes place. Three majors hypotheses are considered as an answer to this question:

a. the self-generation of technology, i.e. technology drives itself through a cumulative and developing dynamic;

b. the economic demand exercised through the market-place, because the technological progress is the result of the economic demand for new products by potential users;

c. the organizational innovations produced by governments, corporations, social movements and universities.

This last argument will be developed, because the rapid growth in new and traditional forms of linkage between industry and universities is the basis of an emergent social paradigm for economic development. Future economic growth is dependent not simply on a new cycle of innovation but on new structure for innovation that ties basic and applied research even more closely together.

We consider that engineering schools represent the primary academic site for relationship with industry. In the "applied" model of university - industry there is a movement from the science to a technology and resulting products. Two scenaria can be considered: in the first, the academic sector came out of the analysis of industry production problems; in the second, the industrial sector arose out of the implications of university research findings. In either format, academic leadership was crucial for both industrial and disciplinary innovations.

So, the academic - industry relations would consists on two linear models going in opposite directions: from product development to academic research and vice-versa. However, according to the considerations on the hierarchical DLs, a spiral model with feedback loops at different points appear more suitable. This spiral model of interaction in both directions, with cooperative arrangements between university and industry has yet to be fully accepted, but it is important to mention that academic - industry relations cannot be seen only in terms of the complex interactions between these two institutional spheres. A third component must be the government actions at various levels, due to their influence in developing strategies and forming new institutions.

3.4 The social entropy and the migration of specialists

Of course, different courses and different ways of development appears in different countries. For example, in the former communist countries of Eastern Europe institutional arrangements, consisted of a series of concentric circles, with government surrounding and industry and the university inside. In this setting there were formal relationships between research institutes and industry for ideological reasons. The intelligentsia were expected to support the working class but in a system largely lacking industrial innovation there was little need for that research. In the special case of Romania, the most embarrassing social phenomenon was the destruction of elites by pushing at maximum the egalitarianism (and so maximizing the social entropy) between social classes. The only aim of the young and performant specialists was to quite the totalitarian state and to reach a free country.

With the collapse of socialism, the institutional sectors are moving apart from each other. Government has reduced its funding of the research institutes. If the research institutes are to survive, they must secure funds from foreign economies or attempt to set up their own industrial sector. Many of the institutes are beginning to develop small "filial firms" that stay for few time within the research institutes before moving out. Its a new way for generating technological change and consequently a new mode of economic development. But the will to migration of the best specialists, especially those of the top sectors (computers, informatics, control engineering) remain unchanged. That continuity is due to a inefficient interaction among universities, industry and government, which have not succeed to find a niche were research can be connected to a local resource (like, for example, applying computer technology to the forestry industry in Finland). That lack of fit generates as feedback another augmentation in the social entropy and produces social friction, professional insatisfaction and minimal economic development, but of course causes problems in other countries because the migration is not acceptable nowhere at a higher rate. The only solution we see is to draw a new social contract between the university and the larger society, in which public funding for the university is made contingent upon a more direct contribution to the economy. To the extent that universities become more entrepreneurial, there will be even more than the traditional tension between research and teaching. As the university crosses traditional boundaries through linkage to industry, it must devise formats to make its multiple purposes compatible with each other. But always the university has been an ingenious innovator and now, mixing disciplinary departments with interdisciplinary centers, a nice perspective of development appears as possible. And, consequently, an embedding of the local social relations which can produce through globalisation a significant support to the international stability.

4. A BETTER WAY FOR INTERNATIONAL STABILITY

In short, we live in this world. We thus have a deep human need to know and to trust it, to be emotionally involved in it. We often have the impression that what remains of the world is a collection of isolated fragments, without any structure and coherence.

It is our conviction that the time has come to make a conscious effort towards the construction of global world views, in order to overcome this situation of fragmentation. But in this case the entropic model seems to poor and can be considered only as a background for more ambitious perspective. In the following such panorama models are named world views. A world view is a coherent collection of concepts and theorems that must allow us to construct a global image of the world, and in this way to understand as many elements of our experience as possible. Societies, as well as individuals, have always contemplated deep questions relating to their being and becoming, and to the being and becoming of the world. The configuration of answers to these questions forms their world view. The project of consciously constructing a world view is indeed an urgent one, since most of the macro-problems and micro-problems of our present time are directly or indirectly related to this situation of fragmentation. It is precisely because we lack such global views of the world that our ability even to start looking for lasting solutions to these problems is limited. We can illustrate the relationship of world views to the current problems facing humanity with two examples.

4.1 The North-South conflict.

In fact, it is a generic name for the conflict between richest and poor countries. Opinions of individuals and groups on this problem are quite divergent. Questions of how to evaluate the level of development of a society, as well as differing visions of possible interactive mechanisms among societies, play crucial roles in the analysis of this situation. As an example, we will consider two sets of opposite views (views A and B, and views 1 and 2), each connected to very different world views.

According to view A, interactions between societies have a destabilizing and degrading effect, due to the irreversibility law. This view would have people strive towards a collection of relatively stable but isolated societies. View B, to the contrary, identifies interactions among societies as enriching and deepening. It therefore aims at a situation of actively interacting societies. Views A and B are intersected by views 1 and 2. View 1 sees the development of societies as mainly characterized by the amount of material and economic prosperity (i.e. the standard of living). Social and cultural developments are believed to be the consequences of this prosperity. According to view 2, on the other hand, the social and cultural development of a society is determinative of its material and economic prosperity.

Variations and combinations of views are possible. Combining view A and view 1 (view A1) would, for example, call for a greater material and economic buffer between the less developed third world and the more developed first world. This would require protectionist measures, and the restriction and manipulation of immigration. Cultural exchanges (tourism, sports, art), however, are not identified as threatening. Views A2 would see a need for greater cultural isolation between North and South, while economic interaction would still be allowed. View B1 would seek more intense economic interaction between North and South, with an eventual unification of economies. The economic model of the North would be the model for this interaction and unification. View B2 would strive towards a greater cultural interaction between North and South, and eventually towards a complete cultural integration. But neither North nor

South would be the model, since neither could be considered to be more developed culturally. Cultural interaction is considered to be an inspiration for a society at large and hence for material and economic prosperity.

It is clear that each of these possible views will lead to different strategies of action. The political consequences for international institutions, such as migration-policy and development-aid are obvious. The search for an answer to this type of question points to the need for knowledge about the mechanisms of resource transfer related in section 2, and for offer a solution to problems that arise in relation to international, inter-economic and inter-cultural relations. The construction of world views is not an easy enterprise. Indeed the fragmentation of our present world has deep structural causes that are ultimately related to the turbulent processes of modernization that society has known during the past centuries. There is today an unmistakable trend towards pluralisation of culture and individualization of human behavior. Instead of one view of the whole, shared by the members of a bounded collectivity, there now exist in our international world very divergent and competitive conceptions and life styles. We do not interact with one culture, but with many cultures, and even with subcultures and fragments of cultures. The individual is forced to select his or her own future life, having to choose between the enormous number of possibilities offered.

4.2 The conflict between specialists and lay - persons.

A second cause of the loss of direction originates from the growing gap between the specialist and the lay person. The rapid development and enormous growth of the sciences, and constant changes in cultural life and in the arts make it impossible for the lay person to keep up. One cannot be a specialist in all domains, and what penetrates into the popular culture from the sciences and the arts is often completely alienated from the specialists' practice. Such a gap exists not only between the specialist and the public, but even within the sciences, as the trend towards specialization in disciplines and sub-disciplines continues to accelerate. The contact between exact sciences and human sciences is scarce and occasional. In recent years, even within isolated disciplines, new boundaries have been created, for example between macro and microbiology, and between macro and micro-economics. In its rapid evolution, science loses the capacity to generate a global view.

World views, as related to the sciences, ethics, arts, politics and religions, are integral parts of all cultures. Moreover, interactions between cultures change constantly. The problem of world views is thus connected to the many attempts at constructing a new coherence between cultural fragments that are constantly emerging and interacting. We believe that it is the task of our time to search for world views in which different systems of interpretation and ideals can be incorporated and can converse with each other. This task is urgent, not only for the multi-cultural societies now found in all major cities of the world, but also for those countries in which a variety of cultural patterns, with quite different histories, are striving towards a certain symbiosis.

From this survey of the forces that stimulate the fragmentation of our knowledge and of the world the difficulty of the task of world view construction emerges. The ultimate goal may not be to simply try to erase any variety of views. The pluralisation of our culture, the depth of the sciences and the arts achieved by specialists and specialization are worthy accomplishments of our culture. We believe, however, that the fragmentation of our culture should be a starting point for a new effort at integration, one that explicitly takes into account these achievements. It is for this reason that the boundaries that have grown between the different isolated fragments must be removed and a new communication among these diverse specialization's must be created. In this direction, the problem of the migration of specialists, which causes a local growth of social entropy, may produces a favorable effect in integration of cultures and so contributes to the growth of the international stability. These two opposite effects must be very well controlled, because it is a fragile equilibrium state and only the politics of globalisation, especially in Europe, can lead to a expected result.

The immense house that our world has become, in which many have lost their way, while others have concentrated so hard on the elaborate construction of just one of its many rooms, needs to be rebuilt and rearranged. We need to make new doors, and create a house where we all can feel at home.

4.3 To a global model based on world views

The project of world-view construction requires that one takes into account subjective as well as objective elements. All knowledge, meaning and value are subjective and bound to a culture insofar as they are necessarily experienced by concrete, historically determined subjects. The nature of this situation does not imply, however, that the objective aspects of a world view are any less important. It is precisely these aspects that are at the basis of the possibility of integration.

Therefore, a world view must not only contain a model of description, an explanation and an evaluation, but also an organized view of the factual and possible influences that humans can have on the world. To define such an integral pattern of action, it is necessary to bring together deeply divided disciplines, and perhaps even to use them for purposes for which they were not originally developed. In this sense, world view construction inevitably has a collective dimension.

The main properties of a world view are coherence and fidelity to experience. A world view cannot contradict known experimental facts, but this does not mean that it coincides with them. In this sense a world view is a continuation of what the sciences pass on to us, sometimes coinciding with it, sometimes generalizing from it, and sometimes critically rejecting it. A world view, however, cannot be determined by its relation to the sciences alone. Our experience also contains our different systems of meaning and our different systems of values. Every experience leads towards action of the one having the experience. It is by means of these actions that we can influence the world, and strive for certain ends. Political praxis, with its many attempts to construct a new society, must also be included in a global world view. Every world view will therefore necessarily contain ideological elements.

The experience of science with its plurality of disciplines, the experience of ethics with its plurality of ethical systems, the experience in aesthetics with its plurality of arts and styles, the experience of politics with its plurality of attempts to construct new societies: all this has a profound influence on world views. The goal is to make the communication between the different layers of our experience explicit.

In the process of world-view construction, the subject does indeed construct a model of reality. Our way of taking into account the presence of the subject in this process is by investigating the process of model construction externally. This requires a paradigm for the process of model construction. To define such an integral pattern of action, it is necessary to bring together deeply divided disciplines, and perhaps even to use them for purposes for which they were not originally developed. For example, the engineer has to organize a production process as a totality that relates purely physical processes with economic, social, psychological and ecological problems. This requires a general theory of problem solving and strategic action, as a frame of reference for all applied sciences.

5. CONCLUSIONS

Quite not very homogenous, the paper use a leading concept - that of social entropy - in order to offer solutions for models that vary from local systems which transfer resource units to global systems which represents world views. Another common reference in these models is the technological change, defined as a principal actor in the evolution of social- economic systems. The following six implications are the most important conclusions of this approach:

First, the problem of having to denote a point of reference for the dynamic analysis of technology is connected to the modes in which evolutionary economics may benefit from sociology through a discourse (translation) or a communication network (stabilization). Second, it follows that the hypothesis of self-organization is crucial for further research, especially based on concepts of variation (as adjustment and reallocation of RTUs) and selection (as instantiation of a virtual system). Third, the discussion points to the importance of sectoral analyses for further elaborating the dynamics of complex systems. More specifically, the idea of a loose coupling of nearly DL may serve to characterize both macro-dynamics structure and micro-economics behavior. Fourth, one might expect policies to be more effective when they support the "creative destruction" of hypercyclic regimes, i.e. the more the system disintegrates, the more active the subsystems can be. In such entropy system, new policies which support emerging trajectories have to be invented. Fifth, the political process has to organize a legitimate intervention. We assume that among the various subsystems, one most accessible to the political process is the dynamics of the social organization and here the role of the universities and of their top products is predominant. Sixth, the local entropic model can be generalized as a regional or quite world view, in search of stability. No priority may be given here to certain partial strategies. Both centralizing and decentralizing forms of organization, both autonomous and hierarchically related structures will appear in various domains, and will have to be evaluated in function of their purposes and efficiency in certain concrete situations.

REFERENCES

Callon, M (1992). The Dynamics of tehno-economics networks, in. *Technological Change and Company Strategies,* (Coombs, R (Ed)), London, p.72-102,

Freese, L (1988). Evolution and Sociogenesis, *Advances in Group Processes*, vol.5, JAI Press, p. 53-118

Georgescu-Roegen, N (1971). *The Entropy Law and the Economic Process*, Cambridge, Harvard University Press

Iacobas, A.D. (1995). Elements of social thermo-dynamics", *Proceedings Symp. on Society Development, (*PHARE for Democracy project.)

Kaneko, K (1991). Climbing up the Dynamical Hierarchy, in Baier, G. and Klein, M. (eds), *A chaotic Hierarchy,* London: World Scientific Publishers, p.235-250

Morrison, F (1991). *The Art of Modeling Hierarchical Systems*, London: John Wiley & Sons

STABILITY OF ECOLOGICAL ECONOMIC LINEAR MODELS (SENSITIVITY ANALYSIS)

Alexander Mironychev

Institute for Systems Analysis
9, Prospect 60 Let Oktyabrya 117312, Moscow, Russia
e-mail: amir@isa.ac.ru

Abstract: New methods of Δ-matrices and δ-vectors are proposed for sensitivity analysis of linear ecological-economic systems. As an application of this technique, a system of models for Russia is considered. It consists of three models: a macro-economic model, an input-output model with 18 sectors, and a pollution model with two pollutants - solid particles and CO2 emission. Sensitivity functions (elasticity) of output and pollution to structural parameter changes are estimated. It was demonstrated that such types of models are sustainable for large and small parameter changes. *Copyright © 1998 IFAC*

Keywords: matrix methods, parameters, sensitivity functions, ecology, economic system

1. INTRODUCTION

One of the important problems of sensitivity theory (Frank , 1978; Kuh et all, 1985) is to estimate changes in endogenous variables of a mathematical model when structural parameters change. As a rule, refined methods are required for investigations of real systems (Alsmiller et all 1983; Schintke, 1976, Elker, 1992; Terlau, 1992). In this article, a new technique of Δ-matrices and δ-vectors is proposed for sensitivity investigation of a linear ecological economic model. In section 2 this new technique of Δ-matrices and δ-vectors is described. Formulas for logarithm sensitivity functions (elasticity) for liner models are derived. Section 2 provides a description of the system of models for Russia. This system consists of a macro-economic model, an input-output model with 18 sectors, and a pollution model with two pollutants - solid particles and CO2 emission. In section 3, the numerical results of the sensitivity analysis are presented. Section 4 is a Summary in which the results obtained are discussed.

2. SOME ELEMENTS OF MATRIX DIFFERENTIATION BY Δ-MATRICES AND δ-VECTORS

In numerous applications, it is necessary to calculate derivatives of complex matrix expressions with respect to elements of the matrices under consideration. For these purposes, it is useful to use the Δ-matrices and δ-vectors technique. The following definitions are necessary.

Definition 1. A vector of dimension n in which all elements are zero but a k-th element is unity is called a δ^k-vector:

$$\delta^k = \begin{bmatrix} 0 \\ \dots \\ 1 \\ \dots \\ 0 \end{bmatrix} k \qquad (1)$$

It is clear that the element s of the δ^k-vector is a Kronecker symbol:

$$\delta_s^k = \begin{cases} 1, & \text{if } k = s \\ 0, & \text{otherwise} \end{cases}$$

Definition 2. A square matrix of dimension n in which all elements are zero but the (i,j)-th element is unity is called Δ^{ij} matrix:

$$\Delta^{ij} = \begin{array}{c} \begin{array}{ccccc} & & j & & \end{array} \\ \begin{bmatrix} 0 & \ldots & \ldots & \ldots & 0 \\ \ldots & \ldots & \ldots & \ldots & \ldots \\ 0 & \ldots & 1 & \ldots & 0 \\ \ldots & \ldots & \ldots & \ldots & \ldots \\ 0 & \ldots & \ldots & \ldots & 0 \end{bmatrix} \begin{array}{c} \\ \\ i \\ \\ \\ \end{array} \end{array} \quad (2)$$

It is clear that the element pq of the Δ^{ij} - matrix vector is also a Kronecker symbol:

$$\Delta_{pq}^{ij} = \begin{cases} 1, & \text{if } i = p \text{ and } q = j \\ 0, & \text{otherwise} \end{cases}$$

The δ-vectors and Δ-matrices help establish the following facts.

Proposition 1. The derivative of the vector $b = \{b_s\}$ with respect to its k=th element is the δ^k-vector.

Proposition 2. The derivative of a matrix $A = \{a_{pq}\}$ with respect to its element (i,j) is the Δ^{ij} - matrix:

$$\partial_{ij} A = \Delta^{ij} \quad (3)$$

The propositions above are easily proven by direct differentiation.

For simplicity of notation, the symbol ∂_k (or ∂_b) will be an operator of the derivative of vector b with respect to its element b_k:

$$\partial_k b = \frac{\partial b}{\partial b_k} \quad (4)$$

And the symbol ∂_{ij} (or ∂_A) will be an operator of the derivative of matrix A with respect to its element a_{ij}:

$$\partial_{ij} A = \frac{\partial A}{\partial a_{ij}} \quad (5)$$

Such notations make it easy find the expression for the derivative of different matrix expressions. Consider the derivative of inverse matrix A^{-1} with respect to element a_{ij} of the matrix A. To prove this, let us denote the element of the inverse matrix A^{-1} as $\hat{a}_{ij}$.

Proposition 3. The derivative of the inverse matrix A^{-1} to the element a_{ij} is:

$$\partial_{ij} A^{-1} = -A^{-1} \Delta^{ij} A^{-1} \quad (6)$$

Proof is easy to obtain by differentiating the equation:

$$A \cdot A^{-1} = E \quad (7)$$

where E is a unity matrix. The differentiation gives the following:

$$\partial_{ij}\left(A \cdot A^{-1}\right) = \partial_{ij} A \cdot A^{-1} + A \cdot \partial_{ij} A^{-1} = 0 \quad (8)$$

Since

$$\partial_{ij} A = \Delta^{ij} \quad (9)$$

it follows that :

$$\partial_{ij} A^{-1} = -A^{-1} \Delta^{ij} A^{-1} \quad (10)$$

For each element $\hat{a}_{pq}$ of the matrix A^{-1} the next expression is valid:

$$\partial_{ij} \hat{a}_{pq} = \hat{a}_{pi} \hat{a}_{jq} \quad (11)$$

Example. Consider the square matrix A of the second order:

$$A = \begin{bmatrix} a_{11} & a_{12} \\ a_{21} & a_{22} \end{bmatrix} \quad (12)$$

Its inverse matrix is:

$$A^{-1} = \frac{1}{detA} \cdot \begin{bmatrix} a_{22} & -a_{12} \\ -a_{21} & a_{11} \end{bmatrix} = \begin{bmatrix} \hat{a}_{11} & \hat{a}_{12} \\ \hat{a}_{21} & \hat{a}_{22} \end{bmatrix}$$

The derivative of A^{-1} to the element a_{21} will be:

$$\partial_{21} A^{-1} = -\begin{bmatrix} \hat{a}_{11} & \hat{a}_{12} \\ \hat{a}_{21} & \hat{a}_{22} \end{bmatrix} \cdot \begin{bmatrix} 0 & 0 \\ 1 & 0 \end{bmatrix} \cdot \begin{bmatrix} \hat{a}_{11} & \hat{a}_{12} \\ \hat{a}_{21} & \hat{a}_{22} \end{bmatrix} =$$

$$= -\begin{bmatrix} \hat{a}_{11} \cdot \hat{a}_{12} & \hat{a}_{12} \cdot \hat{a}_{12} \\ \hat{a}_{11} \cdot \hat{a}_{22} & \hat{a}_{12} \cdot \hat{a}_{22} \end{bmatrix} \quad (13)$$

In many economic applications, the following matrix plays the central role:

$$\hat{A} = (E - A)^{-1} \quad (14)$$

Here E is a unit matrix and A is a so called input-output matrix. This matrix has only nonnegative coefficients. For this matrix, the following result is valid:

Proposition 4. The derivative if matrix $\hat{A}$ with respect to element a_{ij} is:

$$\partial_{ij}\hat{A} = \hat{A}^{-1}\Delta^{ij}\hat{A}^{-1} \qquad (15)$$

The formulae obtained were known earlier (Faddeev and Faddeeva, 1979), but it seems that the technique with Δ-matrices and δ-vectors simplifies calculations and permits to calculate derivatives of arbitrary linear matrix systems. Such a technique is very effective for sensitivity analysis of linear econometric models and input-output models.

3. SENSITIVITY ANALYSIS OF ECOLOGICAL INPUT-OUTPUT MODELS

Let us consider the input-output model with an ecological block (Leontief and Ford. 1972):

$$\begin{aligned} X &= AX + Cq + Y \\ Q &= BX - q \end{aligned} \qquad (16)$$

where X is the vector of industrial outputs. $X = (x_1,\dots x_n)'$; Y is the vector of demand. $Y = (y_1,\dots y_n)'$; Q is the vector of pollution. $Q = (Q_1,\dots Q_m)'$; q is the vector of abatement. $q = (q_1,\dots q_n)'$; A is the input-output matrix. $A = \left[a_{ij}\right]_{j=1,n}^{i=1,n}$; C is the matrix of relative industrial expenses for abatement. $C = \left[c_{kl}\right]_{l=1,m}^{k=1,n}$; B is the matrix of relative industrial pollution. $B = \left[b_{rs}\right]_{s=1,n}^{r=1,m}$.

Here and subsequently it is assumed that during abatement, the pollution is not produced. Below, the Einstein rule for summation will be used. This means that variables with the same indices mean the sum of these variables:

$$a_i b_i = a_1 b_1 + \dots + a_n b_n \qquad (17)$$

This rule will be valid only for variables not separated by parentheses.

The sensitivity investigation consists of estimating the derivatives (sensitivity functions) and elasticity of the industrial pollution Q to structural parameters of matrices A, B, C, and industrial output:

$$\frac{\partial Q}{\partial a_{ij}}, \frac{\partial Q}{\partial b_{ij}}, \frac{\partial Q}{\partial c_{ij}}, \frac{\partial Q}{\partial x_i}$$

As a whole, it is necessary to calculate $N = m\cdot(n^2 + 2mn + n)$ derivatives.

It is more convenient to write our system in the form:

$$\begin{aligned} X &= (E - A)^{-1}(Cq + Y) \\ Q &= BX - q \end{aligned} \qquad (18)$$

or in the form:

$$\begin{aligned} X &= \hat{A}(Cq + Y) \\ Q &= BX - q \end{aligned} \qquad (19)$$

Using the results of section 2, it is easy to find an expression for sensitivity functions. For example, the expression for the derivative of Q to element x_k of the vector X will be:

$$\partial_k Q = B\cdot\delta^k \qquad (20)$$

For coordinates, it will be:

$$\frac{\partial Q_k}{\partial x_j} = b_{ks}\cdot\delta_s^j = b_{kj} \qquad (21)$$

The expression for elasticity Q_k with respect to x_j will be:

$$S(Q_k, x_j) = \frac{x_j}{Q_k}\cdot(b_{kj}) \qquad (22)$$

In this expression, the parentheses mean that we have a product of variables x_j and b_{kj} rather than a sum of them. All sensitivity functions for Q are represented in the Table 1.

4. NUMERICAL SENSITIVITY ANALYSIS OF THE ECOLOGICAL -ECONOMIC MODEL FOR RUSSIA

The technique developed herein was applied to sensitivity analysis of the ecological-economic model for Russia. This model consists of three models: a macro-economic model, an input-output model with 18 industrial sectors, and an ecological block.

The industrial sectors x_i are:

1) Machinery
2) Construction
3) ferrous metals
4) nonferrous metals
5) coal mining
6) oil-
7) other fuel
10) forestry
11) construction materials
12) textiles
13) food
14) other industries
15) agriculture
16) transport& communication

8) electricity
9) chemicals
17) trade
18) other sectors&services

Two types of pollutants were considered: Q_1 is solid particles in the atmosphere and Q_2 is CO2 emission.

Russia and the USA are the two largest in the world. In 1990 they produced 32% of the total CO2emission (Russia, 2,388,720 tho.t; USA. 4,095,022 tho.t). Because of the serious economic crisis in Russia, its GDP has decreased almost two fold since 1990 (see Table 2). For this reason, the emission of pollutants in the air has also decreased. Nevertheless, an exit from crisis and enty into to economic growth will inevitably increase pollution. The problem facing us is to estimate the sensitivity of the pollution to economic activity and structural changes. For the mathematical-computer model under consideration, this means the estimation of sensitivity functions and elasticity of the variables Q1 and Q2 with respect to coefficients of matrices *A*. *B*, *C*, and industrial outputs X_i. The sensitivity analysis of the input-output models without the ecological block were done earlier by Schintke (1976), Kiy (1984), and Lipinski and Gruber (1992).

In Fig.1 the sectoral pollution of solid particles is presented. It is clear that the 'dirtiest' sectors are 8-electricity (103 mln.t); 4- nonferrous metals (64 mln.t); 3- ferrous metals (53 mln.t); 6- gas&oil (37 mln.t); 11- construction materials (20 mln.t).

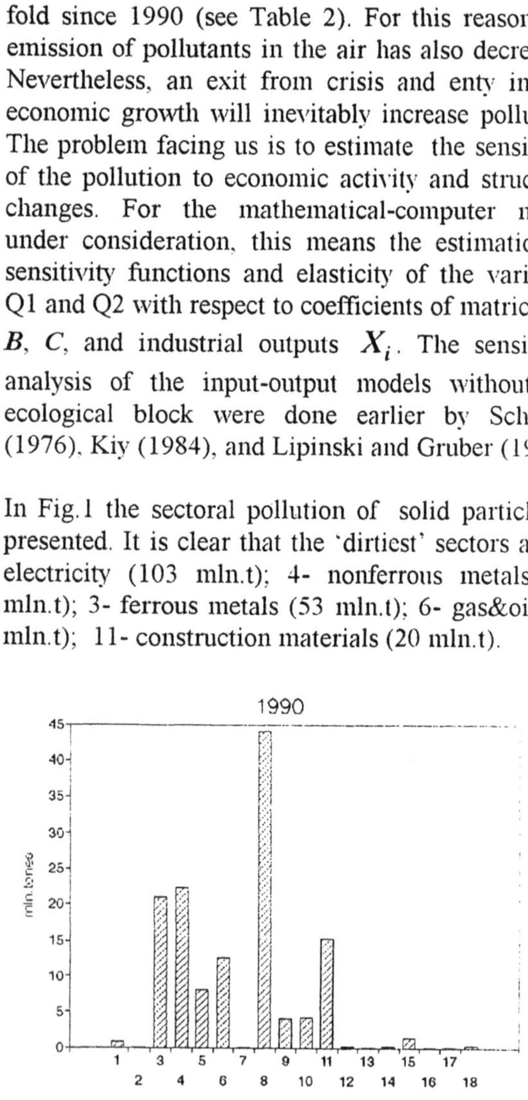

Fig.1. Pollution by sector, Russia, 1990, mln.t.

Fig.1 shows that the share of each sector in total pollution differs significantly. It would be reasonable to expect that the sensitivity of sectors to structural changes will differ, too. Table 3 represents the elasticity of pollutant Q1 with respect to sector outputs *X* and parameters of matrices *B* and *C*.

The following sectors have the highest elasticity with respect to relative abatement expenditures c_j: 2-construction (0.935);, 1- machinery (0.214); 3-ferrous metals (0.134); 14- other industries (0.123); 8- electricity (0.107).

The following sectors have the highest elasticity on relative pollution b_j: 8- electricity (1.08); 4-nonferous metals (0.671); 3- ferrous metals (0.556); 6- oil&gas (0.392); 11- construction materials (0.211).

All these data may are represented in Table 4 in which the five highest values marked by asterisks. It is clear from the table that the sectors with the most influence on the ecological situation are: 8-electricity; 3- ferrous metals; 4- nonferrous metals; 11- construction materials. The most significant sector is 8- electricity. Particularly, the main attention should be paid to parameters $c_{8,1}$ and $b_{1,8}$. The share of other sectors is negligible.

5. CONCLUSIONS

The following conclusions may be drawn:

1. The technique of Δ-matrices and δ-vectors is very effective for sensitivity analysis of large linear systems. It makes possible easy calculation of derivatives of matrices of comlex structures.
2. Derivatives with respect to elements of matrices may be considered as Frechet derivatives of the matrix operator.
3. As a rule, the elasticity under consideration is very small. This means that radical changes in ecology and economics require a long period of time. Single changes (especially in the short run) do not change the picture as a whole. The complex (system-wide) efforts are necessary to changing the situation as a whole.

REFERENCES

Alsmiller R.G. et all (1983). Sensitivity theory and its Application to a Large Energy-Economics Model. *Opers.Res.*, Vol.31, **5**, 915-936.

Elker J. (1992). Forecast and multiplier sensitivity analysis with respect to changes of structural equations. In: Econometric Decision Models. New Methods of Modeling and Applications (J. Gruber. Ed). 502-515. Springer-Verlag, Berlin-New York.

Faddeev D.K., Faddeeva W.W. (1979). Numerische methoden der Linearen Algebra, Munchen.

Frank P.M. (1978). Introduction to SystemSensitivity, Academic Press, New York.

Kiy M. (1984). Ein Disaggregirtes Prognosesystem fuer die Bundesrepublik Deutschland (Lecture Notes Notes in Econometrics and Mathematical

Systems, **224**). Springer-Verlag. Berlin. New York.

Kuh E. Et all (1985). Structural Sensitivity in Econometric Models. Willey Series in Probability and Mathematical Statistics. New York. Toronto.

Leontief W., Ford D. (1972). Air Pollution and Economic Structure: Empirical Results of Input-Output Techniques, 9-30, in: Brody A., and Carter A. (eds.). Input-Output Technique. American Elsiver, New York.

Lipinski C. and Gruber J. (1992). Tolerance Analysis for Structural Parameters in Linear Economic Models. Diskussionbeitrag. **196**. FernUniversitaet Hagen, BRD.

Schintke J. (1976). Sensitivitaetsanalysen im Statischen Offnen Leotief-Model. DIW-Beitrage zur Strukturforschung, Heft 42. Dunker&Humbolt, Berlin.

Terlau W. (1992). Structural sensitivity analysis in econometric models. In: Econometric Decision Models. New Methods of Modeling and Applications (J. Gruber, Ed), 516-536. Springer-Verlag, Berlin-New York.

ACKNOWLEDGMENTS

This article was prepared with financial support of the Russian Foundation for Fundamental Research (Project 96-06-80139). The author also thanks Prof. J.Gruber (FernUniversitaet Hagen), Prof. W. Krelle (Bonn University), Dr. J. Bartnik (University of Frankfurt) for their help and assistance with research.

APPENDIX

Table 1 Derivatives of the output vector X and pollution vector Q with respect to structural parameters, sector outputs and abatement

vector	A	B	C	X
X	$\hat{A}\cdot\Delta_A\cdot\hat{A}\cdot\hat{Y}$	0	$\hat{A}\cdot\Delta_c q$	-
Q	$B\cdot\partial_A X$	$\Delta_B X$	$B\cdot\partial_C X$	$B\cdot\delta_X$

In the table, the following denotations are used:

- $\hat{Y} = Cq + Y$
- Δ_A is a Δ-matrix of the matrix A
- δ_X is a δ-vector of the vector X
- $\partial_A = \dfrac{\partial}{\partial a_{ij}}$ is the derivative operator to a_{ij}
- $\partial_C = \dfrac{\partial}{\partial c_{ij}}$ is the derivative operator to c_{ij}

Table 2 Dynamics of the GDP and pollution in Russia

y-rs	91	92	93	94	95	96
GDP	95	81	74	64	62	59
Q1	31	28	24	22	21	-

Comments: GDP (1990=100); Q1 - thous.t

Sources: Nationalnie cheta Rossii, GOSKOMSTAT, 1997(in Russian); Ohrana Okruzhaushei sredi, GOSKOMSTAT, 1996 (in Russian).

Table 3 Sector indicators: Q1 - pollution (mln.t), X -outputs (bil.rubles), S- elasticity

sectors	Q1	X	S(C)	S(B)
1	1.86	322.48	0.214	0.019
2	0.12	192.65	0.935	0.001
3	53.15	58.87	0.134	0.556
4	64.07	74.21	0.073	0.671
5	16.90	24.57	0.006	0.177
6	37.43	76.35	0.075	0.392
7	0.18	0.48	0.089	0.001
8	103.28	48.78	0.107	1.081
9	11.46	89.58	0.043	0.120
10	5.12	78.00	0.077	0.054
11	20.02	50.14	0.00	0.211
12	0.48	207.16	0.027	0.005
13	0.22	197.42	0.065	0.002
14	0.48	58.10	0.123	0.005
15	2.88	266.60	0.071	0.030
16	0.23	86.40	0.098	0.002
17	0.08	24.58	0.039	0.001
18	0.04	221.09	0.055	0.001

Sources: 'Nationalnie Scheta Rossii 1989-1994', GOSKOMSTAT: Moscow. 1995 (in Russian).

Table 4 Significance of Sectors

sectors	Q1	S(X)	S(B)	S(C	total
1		*		*	2
2				*	1
3	*		*		2
4	*		*		2
5					
6	*		*		2
7					
8	*		*	*	3
9					
10					
11	*		*		2
12		*			1
13		*			1
14				*	1
15		*			1
16					
17					
18		*			1

A SYSTEMS APPROACH TO UNDERSTANDING THE CAUSES OF INSTABILITY IN NATIONS: A CASE STUDY

M.A. Hersh

Centre of Systems and Control and Department of Electronics and Electrical Engineering,
University of Glasgow, Glasgow G12 8LT, Scotland.
Tel: +141 330 4906. Fax: +141 330 6004. Email: m.hersh@elec.gla.ac.uk

Abstract: One useful way of increasing understanding of the causes of violent conflict and possible approaches to its avoidance and resolution is the study of existing and historic conflicts, including comparison of situations which are otherwise relatively similar, except that one is characterised by violent conflict and the other is not. Thus the focus of this work is a comparative case study of the history of Scotland and Northern Ireland from a systems perspective. *Copyright © 1998 IFAC*

Keywords: Historical processes, instability, case study, historical systems.

1. INTRODUCTION: SYSTEMS APPROACHES TO THE STUDY OF HISTORY

Systems approaches have been applied to a range of fields including social and economic ones, but a formalised methodology for the study of historical systems has not yet been developed. The study of history has tended to focus on details and particular significant events, rather than the underlying processes and unifying trends. Classification according to historical periods and natural or political boundaries is clearly convenient, but should not lead to neglect of interactions between periods and across boundaries. Thus history should be seen as process rather than event based. However bifurcation events, where there is a choice of trajectories and the results of the event determines the choice of trajectory, do occur in historical systems, though the result of the bifurcation event may be determined by the previous processes. Historical systems are often highly stochastic, but deterministic historical processes are also important. Significant people and events, such as decisive battles, are more likely to have a significant effect locally than globally, though even at the local level they may purely cause a short lived deviation from the main trajectory. However the presence of such people or events close to threshold phenonoma or chaotic regions may be sufficient to change trajectories or cross boundaries into a region of different systemic behaviour.

Identification and classification of different types of historical processes can lead to greater understanding of the mechanisms and processes which lead to instability and war or a lack of social justice. This could also lead to a unified theory of global history, focusing on underlying similarities rather than the superficial details of events. Values and assumptions are clearly important in modelling historical systems and, for instance, world models have been found to reflect the ideological world views of their creators (Hughes, 1985). In addition the set of possible explanations is restricted by the questions posed (Garfinkel, 1981). Therefore unduly restrictive sets of assumptions or model structure should be avoided and modellers and systems analysts should be aware of their own biases. It has been suggested that non-linear models with both deterministic and chaotic system states could be used to represent peace and war respectively. Control mechanisms could then be applied to avoid the neighbourhood of critical transition points from deterministic regions to chaos and war (Saperstein, 1986). Implementation issues may be even more complex and troublesome in historical than, for instance, industrial systems, and require overcoming political and other resistance to change, as well as determining how best to implement control measures in practice.

Identification of patterns can aid system understanding, but concentration on details can divert attention from underlying relationships. One interesting set of patterns is given by the approximately fifty year Kondratiev cycles. They were initially identified from the analysis of time series data for the period 1790-1920 (Kondratiev, 1924, 1926), though there is some evidence of earlier fifty or fifty-four year cycles (Snyder, 1983). Kondratiev identified four empirical patterns, including increased occurrences of major upheavals, such as revolutions and war, on cycle rising legs. Other evidence for Kondratiev phenomena comes from the ubiquity of cycles and systems theory (de Greene, 1988). Long cycles can be regarded as resulting from the disturbance and restoration of economic equilibrium over a long period and may be a consequence of the drive for continuous economic growth. The time series data on which observation of Kondratiev cycles is based relate to the industrialised countries and the associated theory is concerned solely with capitalist economies. This gives rise to a number of interesting questions as to the nature of systemic behaviour prior to industrialisation and of the less industrialised countries at present. It is also a moot question whether changes in the nature of the economy, for instance in order to prevent environmental crisis, will lead to different types of systemic behaviour, and whether further understanding of the driving mechanisms and links between capitalist economic systems and war events could allow the determination of control inputs to decouple this relationship.

2. A VERY BRIEF DISCUSSION OF SCOTTISH AND IRISH HISTORY

Controlled experiments are frequently used to investigate the satisfaction of hypotheses. However it is generally difficult to obtain suitable control and experimental pairs for historical systems. In addition historical events are generally the result of interconnected factors and processes rather than one single factor. Since history mainly involves the study of past events, it is generally difficult to carry out experiments to test hypotheses. Thus, this approach will be applied here with a certain degree of caution. Northern Ireland and Scotland have been chosen for this study, due to a number of similarities and historical parallels and the presence of stable peace (despite a considerable degree of discontent) in Scotland and a high degree of instability with considerable levels of violence and 'war' deaths in Northern Ireland. The historical sketches given here are necessarily very brief and focus on processes and factors of relevance to stability and instability, while omitting many other important aspects

2.1 Scottish History

By 1035 most of the country was united into a Celtic Gaelic speaking kingdom, though the Hebrides and much of the mainland Highlands were at least in title ruled by Norway until the thirteenth century. Orkney and Shetland only became part of Scotland in 1472. The period between 1035 and the union of crowns in 1603 was marked by conflict between the Scots and English, with England claiming the whole of Scotland and Scotland the three northern English counties of Northumberland, Durham and Cumberland. The most determined attempt to conquer Scotland was made at the end of the thirteenth century and only defeated after many years of fighting, hardship, great brutality and devastation of much of Scotland. However English attempts to subjugate Scotland persisted and the gulf between the Lowland Scots and the Northern English deepened as a result of numerous border raids. Throughout much of this period Scotland was joined in the Auld Alliance with the French which was frequently renewed until the 1550's. It was intended to protect against English aggression, but led to involvement in disastrous military enterprises in support of France and two shattering defeats. Increasing French domination and fear of becoming a French province were central to the 1559-60 reformation/revolution which brought a protestant pro-English regime to power.

In 1603 the crowns of Scotland and England were united, but mutual suspicion prevented union of the kingdoms. Although it was the Scots king who had become king of both realms, London became the centre of affairs with Scotland relegated to a backwater. Anglicisation followed from the royal policy of making the kingdoms similar to each other to reduce conflict. Both kingdoms were protestant, but there were differences in church organisation and forms of worship. The initial religious conflict between catholicism and protestantism gradually changed to one between the Scots presbyterian system and the state. Fears of the imposition of catholicism and opposition to attempts to impose English forms of church organisation were important factors in the resulting war against the king. The Scots invaded England twice, successfully against Charles I and unsuccessfully in support of his son, later Charles II. Scotland then became a conquered province of England for nine years, but this period was marked by orderly government and few abuses. In 1660 control was restored to the Scottish parliament, but the imposition of English forms of church organisation and lay patronage led to dissent and the resignation of 300 ministers. Open air religious meetings were violently disrupted, a peasant rebellion in Galloway was brutally suppressed and the penal acts initially brought in against catholics were used against dissident protestants. The royal prerogative was briefly used (1685-8) to suspend the laws against both catholics and protestant nonconformists. In 1688 the state church of Scotland became presbyterian and patronage was abolished.

In 1707 the Treaty of Union between the parliaments of Scotland and England came into force. It was motivated by fears of catholic heirs to the deposed Stuarts and Scotland siding with England's enemies or choosing a different sovereign. Scotland was generally disadvantaged by it, particularly in parliamentary representation, despite gaining trade with England's colonies, from which it had previously been debarred. The treaty led to rioting in Glasgow and other towns and men started arming and drilling. However armed rebellion did not occur, largely due to lack of unity between the different elements of the opposition and fear of a catholic king. Events such as the abolition of the Scottish Privy council made it increasingly clear that the union was not one of equals.

After being overthrown in 1688 Stuart claimants repeatedly sought to regain the thrones of Scotland and England with the aid of their supporters in the Highland region and at the last attempt in 1745 came within two days march of London before finally being defeated at Culloden. This was followed by brutal suppression. The clan system was broken up by acts forbidding the wearing of tartan and removing the hereditary privilege of Scottish landowners to administer justice on their lands. Clan lands, previously held in common, were transferred to landlords. This led to the clearances in the nineteenth century. These were forced evictions, without compensation and often with great brutality, so that the land could be rented to Lowland sheep farmers and later used for deer stalking. Many communities disappeared as a result. There was also some voluntary emigration. The population of the highlands and islands has fallen from a third to less than a tenth of the total population. The struggle between the Highlands and Lowlands has been a significant part of Scottish history and their relationship has still not been fully resolved.

Between 1889 and 1914 the subject of home rule was brought before parliament thirteen times, receiving a general majority on the last eight occasions and a majority of Scottish members on eleven occasions, but was not allowed to proceed to the committee stage. After the first world war home rule for Scotland ceased to be an important issue in British politics. Both political and cultural, but not paramilitary, nationalist organisations developed. The Scottish National Party (SNP) was founded in 1934 and has concentrated exclusively on constitutional activities and gaining parliamentary representation. A high SNP vote and the shaky position of the labour government led to a referendum for a Scottish assembly in 1979. However it was deemed to have obtained insufficient support, though entry to the European Community was accepted on a lower poll. A further referendum in 1997 for a Scottish parliament with significant tax raising powers obtained a high majority and the parliament will be inaugurated in 2000.

2.2 *Irish History*

By 800 CE there was structural unity in language, law religion and culture, despite little centralised authority. In 1167 the Norman English invaded at the invitation of the King of Leinster and Dublin became the capital of the Norman power in Ireland. Ireland was divided into still unconquered Gaelic areas, feudal liberties with Irish lords owing fealty to the English crown or one of the great English barons, and English land divided into counties and ruled by sheriffs. Prior to the reformation the native Irish were oppressed on the basis of being Irish, with most lacking civil rights in English law and many being reduced to serf or villein status. However attempts to anglicise the country and legislation to outlaw the Irish language and customs were unsuccessful and did not prevent many of the English adopting the Irish language, law and dress.

Subsequent to the Reformation the native Irish and also the Anglo-Norman settlers were discriminated against as catholics. For many years the royal prerogative was used to protect ordinary loyal subjects from religious persecution and to dispense with the oath accepting the sovereign as head of the church of Ireland required of all state and church office holders. Thus the decline in the royal prerogative at the end of the seventeenth century led to a worsening of the position of the catholic majority. Attempts were made to increase the proportion of protestants by plantation with English (and some Scots presbyterian) settlers after confiscation of large estates on various grounds and dispossession of their inhabitants. This was unacceptable even in the sixteenth century and led to considerable resistance. In the mid-sixteenth century the population fell to half a million as the result of ten years of wars and depredations and massacres led to long term resentment. However the only really successful plantation was in Ulster, though a large catholic minority remained. Catholic liberties and religious practices were restricted by a penal code in force from 1695 to 1829 and disabling statutes excluded catholics from office, the army and civil employment. Unlike in England this legislation affected the majority of the population. However the main grievances of the mass of the poor concerned the heavy burden of taxes, tithes to the established church which they did not support, high rents and forced labour on the roads.

From 1714 the government was controlled from Westminster as a subject kingdom in the interests of the protestant ascendancy. The harsh conditions of the peasants gave rise to a widespread agrarian movement and the formation of catholic and protestant peasant organisations. Some concessions were made to catholics, Irish trade and the Irish parliament to gain Irish support in the English conflict with the American colonies (1775-1783). Irish volunteer forces were set up to make up for the absence of the army in America and later attempts to dissolve them were resisted. The Orange order was founded in 1795 to maintain the protestant constitution and soon numbered many thousands of militant protestants. The disarming of Ulster in 1797 involved great excesses and the rebellion degenerated into a war between protestant and catholic militia, though a few years previously catholics and protestants had briefly worked together for reform in the United Irishmen. Ireland became a great wheat growing country following the introduction of legislation restricting its import in 1784, but the staple food was potatoes and the majority were wretchedly poor. Famine followed the potato blight in 1845, resulting in population reduction from eight million in 1845 to 6.5 million in 1848 as a result of famine, fever and emigration. Further decline reduced the population to five million in 1881.

An act of Union was passed in 1800, motivated by fears of the overthrow of the protestant establishment in Ireland and of the consequences of the emancipation of the majority catholic population. Free expression of public opinion was prevented by the presence of 100,000 troops. Promises for full catholic emancipation, state provision for the catholic clergy and settlement of the tithe question took a long time to fulfill. The existing great powers of landlords to eject tenants for arrears of rent were increased. Peaceful meetings and demonstrations of the Catholic Association of Ireland were broken up. North East Ulster began to be associated with a strong anti-catholic sentiment. Rivalries between working class protestants and catholics led to periodic riots in Belfast. The protestants who had opposed union in 1800 supported it in 1900, while the catholic majority was opposed. The Orange order became the strongest bulwark of the protestant anti-home rule cause. At the end of the nineteenth century Home rule for Ireland became a political issue in the British parliament, with the liberals supporting it and the conservatives opposing it. A home rule bill was passed by the commons in 1913, but could not come into force until 1914, at which point the first world war intervened.

Both the nationalist and unionist forces were arming and drilling. A nationalist rising in 1916 was soon crushed, but the execution of fifteen leaders, the imposition of martial law as officials panicked and the internment of thousands more won public sympathy. The six counties of Ulster became a self-governing area in 1920 with their own parliament and government and limited powers, subordinate to Westminster. The Irish Free State (minus these six counties) was established in 1922 as a dominion with full powers of self government and self determination. The Anglo-Irish war from early 1919 to July 1921, euphemistically called the troubles, seriously embittered Anglo-Irish relations. The fact that nationalist support was drawn exclusively from catholics led many protestant unionists to identify catholicism with hostility to the state. Persistent depression and unemployment in the thirties led to a resurgence of sectarian strife and a campaign of violence in the forties. Catholics continued to suffer from discrimination in jobs and housing. There was a further campaign of violence from 1956 to 1962 and nationalist demonstrations were sometimes broken up by the police.

Despite the legacy of resentment and mistrust, in the early sixties the forces for conciliation seemed as strong as those for conflict and the only sizable group opposed to conciliation was the right wing of the protestant community. However disorder mounted and in 1969 the government requested British troops to come in to restore order. Packages of reform measures were introduced too late for catholic gratitude and gave rise to protestant resentment. The strongly reinforced army was involved in confrontations with catholic rioters and the initial goodwill turned to hostility. An IRA offensive followed and was countered by large scale detention without trial, which led to demonstrations and resentment. Direct rule from Westminster was temporarily introduced, but violence increased and protestant private armies emerged. In 1973 a joint administration was set up of moderate groups on both sides and internment ended. Violence declined, but did not cease and has continued to the present. There are now formal structures for consultation with the Irish government and an agreement has recently been concluded, despite frequent disruption of the peace process by outbreaks of sectarian violence.

3. ANALYSIS AND DISCUSSION

3.1 Similarities and Differences

Both populations are mixtures of Celtic and other elements and until relatively recently spoke (different versions of) Gaelic, which has now been almost totally replaced by English. Scotland and Ireland are comparable in surface area and population and have much smaller populations than England.

In both cases the conflict with England has been one of the most significant historical processes. The Scots were more successful at resisting English conquest and Scotland retained its independence several centuries longer than Ireland. This is, at least in part, due to the different political structures in the two countries, with Scotland already united into one kingdom at the start of English attempts at conquest, whereas Ireland was still divided into 5 kingdoms, with a relatively weak high king, though cohesive cultural and legal systems. In the Scottish case the union of crowns was voluntary, with the Scots (initially) retaining the right to choose their own monarch, whereas in the Irish one it was the result of coercion and conquest and resulted in the English sovereign automatically becoming the Irish one.

Religious conflicts have played a significant role in both countries, starting with the conflicts between the Celtic and Roman churches. There is also a history of discrimination and restriction of the civil liberties of both catholics and, to a lesser extent, presbyterians in both countries. There is still some sectarian division, with, for instance, some separate schools in parts of Scotland, but, unlike in Ireland, this has never been the cause of significant conflict. The presence of a majority protestant community in the six counties of Ulster is the result of a deliberate policy of colonisation and plantation of protestants (both English anglicans and Scots presbyterians), which has no parallel in Scotland, just as there is no Irish parallel to the Lowland/Highland division.

3.2 The Conflict with England

In Scotland violent resistance to English rule largely ceased in 1745, due to a decline in support for the Stuart monarchy, changes in the issues of concern, brutal suppression and depopulation of the Highlands where most support was based, as a result of the clearances, emigration and economic factors. Some low level violent conflict occurred, in response to, for instance, the clearances of the nineteenth century and the rent strikes of the 1920's, but did not last. Paramilitary organisations have not developed in Scotland in support of devolution or independence, whereas there is a history of paramilitary groups in both communities in Northern Ireland and sporadic or more frequent outbreaks of violent conflict.

In Scotland, although the conflict with England has not been satisfactorily resolved, it has become less important and the likelihood of significant violence almost totally disappeared. However in Ireland the conflict has been kept alive, largely through the change of focus to the sectarian protestant and catholic communities. This change is largely a consequence of the power struggles resulting from the reformation and the deliberate 'plantation' of protestants to change the population balance and strengthen English power and control. In Ulster this led to sectarian division between protestant majority and significant minority catholic communities.

Although the communities are almost totally segregated within a given centre of population, areas in which the two communities are found are totally intermixed. Therefore resolution of the conflict by partition is not possible, as illustrated by the failure of the original partition into the Irish Free State (1922) and Northern Ireland (1920) to end the conflict.

3.3 Religion and Ideology and Power

Sectarianism and discrimination have been important in both countries. In Ireland discrimination was originally based on Irish identity, language and customs and intended to prevent assimilation of the Anglo-Norman settlers and adoption of the Irish language and customs. This was supplemented after the reformation by discrimination against catholics. This used the religious and political conflicts of the reformation and the resulting religious differences between Ireland and England to bolster the power of England in Ireland by restricting the political and other liberties of the Irish as catholics.

This is an example of the use of religion as a political control mechanism, with political loyalty requiring religious orthodoxy and uniform religious practice, typical of the post-reformation period. This combines with its role as a source of genuine, albeit dogmatic, belief, leading to strong fears of the imposition of alien forms of worship. The role of religion as a control mechanism, combined with dogmatism and lack of tolerance for diversity frequently made it a force for instability and divisiveness rather than cohesion within and across boundaries. The role of religion has often been taken over by political, economic and other ideologies. Low tolerance of diverse views and the use of ideologies as political or social control mechanisms. frequently lead to instability.

Since one of the roles of religion in the post-reformation period was defining loyalty to the state, it is not surprising that religious differences in England, Scotland and Ireland formed part of the conflicts between them. Differences in power structures are just as important as differences in liturgy and church organisation in determining the differences between the three religions. Specifically anglicanism gave power to the (English) state, presbyterianism gave power, in theory, to the community (the Scottish people) and, in practice, to the more influential, male members of it and catholicism gave power to a distant power/protector (the pope and the church hierarchy in Rome).

3.4 Processes Webs and Bifurcation Events

As already indicated, studies of history frequently focus on specific events (and people). These are often bifurcation events in which two (or more) trajectories are possible following a specific event and choice of trajectory is determined by the result of this event. However the results of an apparent bifurcation event are frequently determined by a range of other factors and the evolution of a number of (possibly interacting) historical processes. In some cases the choice of trajectory determined by the bifurcation event will only be significant in the short term, but have little or no effect in the longer term. For any point in time and space there is a process web of (branching) historical processes which evolve backwards and forwards in time and sideways in space. The concept of process web can be very useful in representing and analysing sequences of historical events and, in particular, in investigating and representing causality and consequences.

3.5 Example: The Norman Conquest of Ireland

The Norman invasion and conquest of Ireland may seem to result purely from the King of Leinster's invitation to the king of England, but at least three different processes are involved. One is the process of political organisation of the country which led to a number of kings rather than a centralised authority, with uneasy relationships, shifting balances of power and feuding between them and a lack of peaceful mechanisms for solving disputes. Another set of processes, including geographic proximity, the demands of the Norman lords for additional land holdings and the desire of the monarch to reward or distance them, led to the Norman conquest. This resulted in an expansionist militarised feudal culture, interested in conquest, in England. A third set of processes relates to the history of the church in Ireland, the conflicts between the Celtic and Roman churches and the subsequent evolution of the Irish church. This led the pope to confer the lordship of Ireland on the English king at the end of the twelfth century and resulted in the submission of a number of Irish religious and secular leaders.

After the reformation there was deliberate 'plantation' of protestants (which was only successful in Ulster) to increase English power and reduce Irish resistance. The previous discrimination against the Irish as Irish was supplemented by anti-catholic discrimination and measures to encourage separation of the planted protestant communities. This also affected descendants of the Anglo-Normans who remained catholic. However, whereas the Anglo-Normans were integrated into Irish society and adopted the Irish culture and language, the catholic and protestant communities remained separate. The resulting sectarian communities in Northern Ireland are one of the main factors in the present instability.

4. CONCLUSIONS

A case study has been carried out of Scotland and Northern Ireland to investigate the reasons for stability in the former and instability in the latter. These two countries were chosen due to the similarities and parallels between their histories. In terms of the three stage model of war (Hersh, 1997) this study shows the importance of creating a climate that does not favour war and mechanisms to resolve conflict. It also illustrates the potentially divisive and destabilising force of dogmatic ideologies which are intolerant of difference.

The conflict in Northern Ireland is in many ways typical of the type of conflict that can occur when two separate communities are forced to live together, though it is unusual in the artificial creation of this situation through a process of 'plantation'. The conflict in Northern Ireland is often labelled as religious, but it is the existence of two sectarian communities with different experiences, access to power and levels of privilege, with one suffering considerable discrimination, that is significant. A similar type of conflict could occur between two communities divided by ethnicity.

The study of history frequently focuses on bifurcation events which determine the subsequent choice of trajectory. However the results of bifurcation events often depend on previous historical processes. The concept of process web of branching historical processes or trajectories has been defined. Process webs can be used to represent and analyse historical systems and, in particular, to investigate and represent causality and consequences. Future work will further develop the methods obtained here and investigate the application of mathematical and control techniques, such as Petri nets, to the representation and analysis of historical processes.

REFERENCES

De Greene, K.B. (1988). The Kondratiev phenomenon: a systems perspective. *Systems research*, vol 5, pp. 281-298.

Garfinkel, A. (1981). *Forms of explanation*, Yale university press.

Hersh, M.A. (1997). A systems approach to the promotion of international stability through the resolution of intranational conflicts, *Transition '97*, Warsaw, Poland.

Hughes, B. (1985). World models. *Int. studies quarterly*, vol 29, pp. 77-101.

Kondratieff, N.D. (1926). Die langen Wellen der Konjunktur. *Arch. Sozialwiss. Sozialpolitik*, vol LVI, pp. 573-609.

Kondratieff, N.D. (1984). *The long wave cycle*, Richardson and Snyder.

Saperstein, A. (1986). Chaos and determinism: a paradigm. *IEEE AES magazine*, June, pp.18-24.

Snyder J.M. (1983). Introduction, Kondratieff, 1984.

Acknowledgement: Fred Kile for helpful comments.

A DYNAMIC-GAME MODEL OF COOPERATION IN ENERGY AND CLIMATE CHANGE

Stefan Pickl, Jürgen Scheffran*

**Department of Mathematics, Darmstadt University of Technology, Germany*

Abstract: To identify, assess and compare options for avoiding anthropogenic climate change and achieving a sustainable energy consumption, the interaction between energy-related carbon emissions, its associated environmental impact and economic growth is analyzed within the framework of a dynamic game model. Basic variables are energy production, emissions into the enviroment, the energy price and the economic output (wealth). Major control parameters are the allocation of funding with regard to various energy options and the degree of international cooperation through technology transfer and capital flow. Conditions are given for Nash equilibria and optimal cost allocation in a game between industrialized and developing countries. Costs of unilateral action can be compared with minimal costs in case of cooperation (in particular, Joint Implementation), by variation of the allocation preferences. The cost savings by the coalition-forming-process could be transfered into a fund and be used to control the process towards a stable equilibrium. To achieve a feasible control set and a control vector, the core and the tau-value of the dynamic game can be determined. *Copyright © 1998 IFAC*

Keywords: Conflict, control, cooperation, dynamic modelling, energy distribution, environment, game theory, resource allocation, stability

1. DYNAMIC-GAME MODEL IN ENERGY AND CLIMATE CHANGE

To understand options for control and conditions for cooperation in energy and climate change, a dynamic game model is applied. The general framework model (called the SCX model) describes the interaction between goal functions $S_i(\mathbf{X})$ of a state vector $\mathbf{X}$, and means C_i to induce changes $\Delta\ \mathbf{X}$ for actors $i = 1, ..., n$. Methods of optimal control and game theory define conditions for cost minimization and cooperation, in particular of Nash equilibria and Pareto optima, depending on the preferences f_{ki} of actors i for distribution of means to options $k = 1, ..., m$ in the action space.[1]

1.1. *Energy Emissions, Goals and Costs*

To analyze the action space in climate change, the model is specified to describe the interaction between two actors ($i = 1$: industrialized country, $i = 2$: developing country) who have the option to invest means (financial costs) $C_i = c_i\,\Delta X_i$ into the installation of additional energy power ΔX_i at unit costs c_i. Both actors can distribute their means to two energy paths, using established energy technology on the one hand ($k = 1$) and more modern energy technology ($k = 2$) on the other hand. $0 \leq f_i := f_{2i} \leq 1$ is the fraction of means of actor i distributed to option $k = 2$. g_{ki} denotes the mean emissions of environmental harmful gases (in particular greenhouse gases like CO_2) per energy unit on the energy path $k = 1, 2$, with $g_{2i} < g_{1i}$, while w_{ki} is the additional wealth associated with an energy unit on path k. Both actors pursue two conflicting goals: a required wealth production $S_i^W \leq \Delta W_i$ (additional economic output) and a required emission change $S_i^G \geq \Delta E_i^G$ for energy-related gases. To achieve both goals, actors can adapt their energy investments and their distribution preferences f_i for the new energy path. In addition, the industrialized country may have the option to invest a fraction $p_1 C_1$ of its investment costs into modern technologies in the developing country to reduce emissions, which are

[1] For an outline of the SCX model with references to earlier work see Scheffran (1996). A first application to climate change is given by Scheffran and Jathe (1996) and an adaptation to Joint Implementation is described in Scheffran and Pickl 1997.

accounted to the industrialized country with percentage h_1 as its own emission reductions (called Joint Implementation). For the case that the industrialized country distributes all its investment means into the modern energy path ($f_1 = 1$), while the developing country initially distributes all its means only to the old energy path ($f_2 = 0$), then the total emission reductions for both actors can be presented as

$$\begin{aligned}\Delta E_1^G &= sc_{11}C_1 + sc_{12}C_2 + G_1^W \\ \Delta E_2^G &= sc_{22}C_2 + sc_{21}C_1 + G_2^W\end{aligned}$$

with
$sc_{11} = (1-p_1)w_{21}^c \Delta g_{21}^w + h_1 p_1 w_{22}^c \Delta g_{22}^w$
$sc_{12} = h_1 f_2 w_{22}^c \Delta g_{22}^w$
$sc_{22} = f_2\, w_{22}^c \Delta g_{22}^w$
$sc_{21} = p_1 w_{22}^c \Delta g_{22}^w$

$\Delta E_i^G = E_i^G(t+1) - E_i^G(t)$: emission change of actor $i = 1,2$ between two time steps
$G_i^W = g_{1i}^w S_i^W$ $(i = 1,2)$: wealth induced emissions for actor i on energy path $k = 1$
$w_{ki}^c = \frac{w_{ki}}{c_{ki}}$: wealth for actor i created per means distributed to energy path k
$g_{ki}^w = \frac{g_{ki}}{w_{ki}}$: emissions by actor i created per wealth unit of actor i on energy path k
$\Delta g_{2i}^w = g_{2i}^w - g_{1i}^w$: gain in emissions per wealth unit for energy path 2 compared to path 1.

1.2. *Stability and Distribution Preferences*

These equations now have the form of the first set of equations of the SCX model. The coefficients sc_{ij}, representing the impact of the energy costs C_j on the emission reduction ΔE_i^G, determine whether stable Nash equilibria exist in the (C_1, C_2)-space. The limit condition for stability is given by the ***stability index***

$$Z = sc_{11}sc_{22} - sc_{12}sc_{21} = 0$$

Since the coefficients sc_{ij} depend on the distribution p_1 of the industrialized country for Joint Implementation and the fraction f_2 of the developing country's investment into the modern energy path, conditions for these preferences can be defined under which both actors can achieve their goals at minimal costs C_i, including cooperation.

It can be shown by game-theoretic considerations that both actors, aiming at pursuing their goals at minimal costs, are tempted to change their preferences at a *decision point* $p_1 = p_1^*$ and $f_2 = f_2^*$, which depends only on the values of the stability index $Z(z_1, f_2)$ for the cases that actors 1 and 2 choose one of the pure distribution preferences $p_1 = 0$ or $p_1 = 1$ and $f_2 = 0$ or $f_2 = 1$ (boundaries of the preference space). The decision point (p_1^*, f_2^*) is a saddle point in Z, and by moving through this point, actors change their preference for one of the two energy paths in order to achieve minimal Nash equilibrium costs (Jathe etal., 1997):

$$\begin{aligned}\widehat{C}_1 &= \frac{sc_{22}(S_1^G - G_1^W) - sc_{12}(S_2^G - G_2^W)}{Z} \\ \widehat{C}_2 &= \frac{sc_{11}(S_2^G - G_2^W) - sc_{21}(S_1^G - G_1^W)}{Z}\end{aligned}$$

For the described allocation problem, actor 1 would reduce its Nash-equilibrium costs $\widehat{C}_1 > 0$ for given goals S_1^W and S_1^G within the preference region $0 \le p_1 \le p_1^* = 1$. Actor 2 would reduce costs within the preference region $f_2 > f_2^* = 0$.

2. COOPERATION IN EMISSION CONTROL

2.1. *Framework for Joint-Implementation*

For adaption to the special case of Joint-Implementation the terminology of the general SCX-Model is varied, using emissions as goals S, money as means C and technology as system variables X. Following this terminology, the TEM-model describes the economic interaction between several actors (players) $i = 1, \ldots, n$ which intend to minimize their emissions (E_i) caused by technologies (T_i) by means of expenditures of money (M_i) or financial means, respectively. The players are linked by technical cooperation and the market, which expresses itself in the nonlinear time-discrete dynamics. The aim is to reach a steady state of the system by choosing the control parameters such that the emissions of each player become minimized. The focal point is the realization of the necessary optimal control parameters via a cost game, which is determined by the way of cooperation of the actors. In application to the work of Leitmann (1971), but not regarding solution sets as feasible sets, the τ-value of Tijs (1981) is taken as a control parameter. This leads to a new class of problems in the area of 1-convex games. In the following, the problem is solved for a special case. With this solution a reasonable model for a Joint-Implementation process is developed, where its necessary funding is represented by the non-empty core of the analyzed game.

$$\begin{aligned}\Delta E_i &= \sum_{j=1}^{n} em_{ij}M_j \qquad (1)\\ \Delta M_i &= -k_iM_i[M_i^* - M_i]\{E_i + \tau_i\Delta E_i\}\end{aligned}$$

E_i Emissions of actor i

M_i Financial means of actor i

em_{ij} effect on emissions of actor i if actor j invests into his technologies

with $E_i = E_i(t), M_i = M_i(t), em_{ij} = em_{ij}(t)$
$\Delta E_i(t) = E_i(t+1) - E_i(t)$,
$\Delta M_i(t) = M_i(t+1) - M_i(t)$ $(i,j = 1,2)$.

2.2. *The Eigen-Values of the TEM-Model*

With $\sum_{j=1}^{n} em_{ij}(t)\hat{M}_j(t) = 0$
and $\hat{M}_i(t)[M_i^* - \hat{M}_i(t)]\hat{E}_i(t) = 0$
we are able to determine the fixed points of the dynamical system (1). If we regard the Jacobi-matrix for the special case $em_{ij}(t) = em_{ij}^*$, i.e. the economical relationships are constant over a long period, we get

$$\begin{pmatrix} 1 & 0 & \dots & 0 & em_{11} & \dots & em_{1n}\\ 0 & 1 & & \vdots & em_{21} & \dots & \dots\\ \vdots & & \ddots & & & & \\ 0 & 0 & 0 & 1 & em_{n1} & \dots & em_{nn}\\ 0 & 0 & 0 & 0 & 1-k_1M_1^*\hat{E}_i & & \\ \vdots & \vdots & \vdots & \vdots & & & \\ & & & & & \ddots & \\ 0 & \dots & & 0 & & & 1-k_nM_n^*\hat{E}_n \end{pmatrix}$$

and the *Eigen-values*: $\lambda_1 = \ldots = \lambda_n = 1$ and $\lambda_{n+j} = 1 - k_jM_j^*\hat{E}_j$ for $j = 1,\ldots,n$.

We see that the fixed points are not attractive. In Krabs and Pickl (1997), the necessary control parameters are developed. The realization of the control set via a played cost game is the main part of the next section. In the last section it is shown, that for a special class of games it is possible to control the system with the τ-value, which was introduced in Tijs (1981).

2.3. *The Cost-Game in the TEM-Model*

The nonlinear time-discrete dynamic of the TEM-model can be formulated as

$$\begin{aligned}\Delta E_i(t) &= \sum_{j=1}^{n} em_{ij}(t)M_j(t) \qquad (2)\\ \Delta M_i(t) &= -k_iM_i(t)[M_i^* - M_i(t)]\{E_i(t) +\\ &\quad +\tau_i\sum_{j=1}^{n} em_{ij}(t)M_j(t)\}\end{aligned}$$

In order to reach steady states, which are determined in Krabs and Pickl (1997), an independent institution may influence the trade relations between the actors. In practice, the imposing of taxes or the giving of incentives means, that in the TEM-model the *em*-parameter will change. Now, the principle of Joint-Implementation implies that the cooperation will be beneficial:

$$\begin{pmatrix} em_{11} & em_{12}+\epsilon & em_{13}\\ em_{21}+\epsilon & em_{22} & em_{23}\\ em_{31} & em_{32} & em_{33}\end{pmatrix}$$

Actor 1 and Actor 2 do cooperate

$$\begin{pmatrix} em_{11}+\omega & em_{12}+\omega & em_{13}+\omega\\ em_{21}+\omega & em_{22}+\omega & em_{23}+\omega\\ em_{31}+\omega & em_{32}+\omega & em_{33}+\omega\end{pmatrix}$$

All players do cooperate

For $K_i(t)\tau_iM_i(t) \geq 0$ the difference between the cooperative and the non-cooperative case is always positive. So we have constructed a reasonable cost-game. Now, the method is that at each time step, this amount is put into a central fund. We determine the core for this set and with the τ-value, which has to lie in it, we want to steer the system. Let us state

Lemma 1: *The game which is defined by (2) is zero-normalized.*

Lemma 2: The 3-person game is super-additive, if $\omega \geq \max\{\epsilon,\delta,\gamma\}$.[2]

Considering that the core is the convex hull of its extreme points we note that the core is a polytope and compact. If we assume that the multistage process is continuous we can formulate the general model for the nonlinear time-discrete dynamic of the TEM-model:

$$\begin{aligned} x_i(t) &\in X_i \subset \Re_j^n\\ u_i(t) &\in \mathbf{Core}(x(t)) \subset \Re_j^m\\ x_i(t+1) &= x_i(t) + f_i(x(t),u(t))\\ x_{t+1} &= T_t[x(t),u(t)]\end{aligned}$$

Then we can proof that

[2] δ is the benefit of the technical cooperation between player one and three, γ between player two and three, respectively.

Theorem 1: *Let $U_i(x_i)$ be a continuous set function for $i = 1, \ldots, n$. The feasible set of control parameters is defined by the core. Then there exists a solution of our problem, which can be determined by solving the Bellman Functional Equations.*

Let us use the following abbreviations for the game

$$v^*(\omega) := v(123)(\omega) \qquad v^*(\epsilon) := v(12)(\epsilon)$$
$$v^*(\delta) := v(13)(\delta) \qquad v^*(\gamma) := v(23)(\gamma)$$

Then we can proof the following theorem

Theorem 2: *If*

$$v^* = \max\{v^*(\epsilon), v^*(\delta), v^*(\gamma)\}$$
$$= \min\{v^*(\epsilon) + v^*(\delta), v^*(\epsilon) + v^*(\gamma), v^*(\gamma) + v^*(\delta)\}$$

then the game in (2) is 1-convex.

2.4. *The equivalence theorem - The existence of the τ-value*

Under consideration that the Theorem 2 is valid, we are able to determine the τ-value, which is demanded to be equivalent to the general control vector $(u_1, u_2, u_3)^T$ or $(u, u+a, u+b)^T, a \in \Re, b \in \Re$, respectively: If game v^* is 1-convex, yields

$$\tau^*(v^*) = b^{v^*} - \frac{1}{n}g(N) = \begin{pmatrix} u_1 \\ u_2 \\ u_3 \end{pmatrix}$$
$$= \begin{pmatrix} v^*(\omega) - v^*(\gamma) - \frac{1}{3}g^*(\omega) \\ v^*(\omega) - v^*(\delta) - \frac{1}{3}g^*(\omega) \\ v^*(\omega) - v^*(\epsilon) - \frac{1}{3}g^*(\omega) \end{pmatrix}$$
$$= \begin{pmatrix} \frac{1}{3}v^*(\omega) + \frac{1}{3}v^*(\epsilon) + \frac{1}{3}v^*(\delta) - \frac{2}{3}v^*(\gamma) \\ \frac{1}{3}v^*(\omega) + \frac{1}{3}v^*(\epsilon) + \frac{1}{3}v^*(\gamma) - \frac{2}{3}v^*(\delta) \\ \frac{1}{3}v^*(\omega) + \frac{1}{3}v^*(\delta) + \frac{1}{3}v^*(\gamma) - \frac{2}{3}v^*(\epsilon) \end{pmatrix}$$

This leads to the following system of equations (3) which we have to solve

$$\begin{pmatrix} \max\{x,y,z\} \\ \max\{x,y,z\} \\ \max\{x,y,z\} \end{pmatrix} + \begin{pmatrix} 1 & 1 & -2 \\ 1 & -2 & 1 \\ -2 & 1 & 1 \end{pmatrix} \begin{pmatrix} x \\ y \\ z \end{pmatrix} = \begin{pmatrix} 3u_1 \\ 3u_2 \\ 3u_3 \end{pmatrix}$$

Theorem 3 (Equivalence -IANUS- Theorem): *The system (3) has got a solution if one of the following cases occur:*

$a = 0$	and	$b \in \Re$,
$b = 0$	and	$a \in \Re$,
$a = b$	where	$a \in \Re$.

It is possible to show that for $(1, \frac{3}{2}, 2)u, u \in \Re$ there always exists an *equivalent 1-convex game*. This is a new result and can be used in a constructive way to realize a Joint-Implementation Program.

REFERENCES

Driessen, T. (1988). *Cooperative Games, Solutions and Applications*, Kluwer Academic Publisher, Rotterdam 1988

Jathe, M.; Krabs, W. and Scheffran, J. (1997). Control and Game-Theoretic Treatment of a Cost-Security Model for Disarmament. *Mathematical Methods in the Applied Sciences* **20**, 653-666.

Krabs, W. and Pickl, S. (1997). Time discrete dynamical games. to appear in: *Mathematical Journal of Optimization*

Leitmann, G. (1971). *Cooperative and noncooperative many players differential games*, Springer Verlag, Wien, New York 1971.

Scheffran, J. and Jathe, M. (1996). Modelling the Impact of the Greenhouse Effect on International Stability. In: *Supplemental Ways for Improving International Stability 1995* (Kopacek, P. (Ed.)), Pergamon, Amsterdam.

Scheffran, J. (1996). Modelling Environmental Conflicts and International Stability. In: *Models for Security Policy in the Post-Cold War Era* (Huber, R.K. and Avenhaus, R. (Eds.)) 201-220, Nomos, Baden-Baden.

Scheffran, J. and Pickl, S. (1997). Control and Game-Theoretic Assessment of Climate Change - Options for Joint Implementation, Paper presented at *International Conference on Transition to Advanced Market Institutions and Economies*, Warsaw, June 18-21.

Tijs, S.H. (1981). Bounds for the core and the τ-value. In: *Game Theory and Mathematical Economics* (Moeschlin, O. and Pallaschke, D. (Eds.)), North-Holland Publishing Company, Amsterdam.

OBJECT ORIENTED MODEL TO SUPPORT EARLY WARNING AND RAPID REACTION PLANNING IN INTERNATIONAL CONTEXT

Velizar M. Shalamanov[(1)] and Todor D. Tagarev[(2)]

[(1)] MoD Research Institute, Sofia, Bulgaria
[(2)] Space Research Institute, Bulgarian Academy of Sciences
6 Moskovska Str., POBox 799, Sofia 1000, Bulgaria
E-mail: tagarev@bgcict.acad.bg

Abstract: The current security environment allows for greater integration of efforts to prevent war and conflict. This paper is focused on the technological aspects of a concept of an international Network of Situation Centers to support early warning and rapid reaction of the international community. It presents the multilayer architecture of the system and object-oriented description of its components. Simulation experience and potential deterring effects of the concept are briefly described.

Keywords: Complex systems, object-modelling techniques, monitoring.

1. INTRODUCTION

The changes in international security environment after the Cold War and proliferation of conflict triggered a new surge of investigations of requirements of crisis management and conflict prevention to support Dynamic Peace (Mueller and Law, 1995; Kile, 1996). For countries with no sufficient resources for large scale military operations, success in Early Warning (EW) of threats and provision of rapid reaction (RR) of the international security system to protect their vital interests becomes a strategic goal. These countries need to precisely formulate their own interests and to legitimate these interests in the international community. Further, they need to define and assess possible threats and to convince others that these threats are threats to international stability.

The latter requires a well developed system of international consultations. From a different point of view, the monitoring of the security situation for the purposes of early warning is a task that calls for all-source data fusion, analysis, assessment and decision making support.

At the same time, development of UN mechanisms and especially the institute of military operations other than war (MOOTW) permit to change the global strategy - preventing large wars is viewed as capability to win MOOTW in proper time and place in order to prevent wars - wars that cannot be won without significant harm to international peace and stability.

Furthermore, early warning without diversified capabilities for adequate reaction adds little to the security guarantees of smaller countries with no defense treaties with a major power or alliance. Rapid reaction needs political will but also well prepared force building blocks, capability based planning system and a deployable command and control (C2) system for synthesized rapid reaction forces.

Analysis of all political, economic, military and other aspects of early warning and rapid reaction (EW&RR) problem is far beyond the goals of this paper. Its focus is on formulation of a project "*EW&RR planning and execution control system*" and discussion of its feasibility. This project is an opportunity for political and economics leadership for commitment to reengineer our world (UN) security and development system for the implementation of basic values and ideas of stability trough security and development supported by an adequate information system.

2. FOUNDATIONS OF THE EW&RR SYSTEM

The technology to build early warning and rapid reaction planning and execution control (EW&RR) system is available. What is needed is political will to implement this technology, to build the system, to keep it operational and to use it in accordance with the UN Charter. There is experience in solving political and financial issues related to GPS, INMARSAT, IRIDIUM, and other large international projects. Also available are two additional tools - networks of non-governmental organizations, willing to promote international stability, and Internet as a medium for real global connectivity.

From a technological point of view, this system can be based on integrated and distributed information environment that supports the following services allocated to eight key servers (Hayes-Roth and Erman, 1994):

Comm server - brokers limited communications bandwidth to consumer applications via contracts for latency and quality of service priced to balance supply and demand;

Data server - employs a common object-oriented command & control schema to provide its clients periodically updated query-based views of distributed, heterogeneous databases;

Web server - provides its clients means to construct, distribute, view, edit, and replicate node-link structures incorporating objects of arbitrary types;

Situation server - enables its clients to develop interpretations or "pictures" of the international system, incorporating objects, aggregates, inferences, and predictions, all of which are indexed over space, time, and assumed context;

Plan server - enables a group of distributed planners to hypothesize, evaluate, and disseminate alternative courses of actions (COAs);

Model server - sets up and executes simulations to assess alternative plans in the context of various assumed situations;

Map server - constructs and renders "maps," which are selected subwebs (including situations and plans), using appropriate symbologies and geospatial registration;

Message server - provides user applications a single interface for accessing, composing, editing, routing, and dispatching military messages.

The servers interact using the Object Management Group's Common Object Request Broker Architecture (CORBA). However, the ORBs conceived of via OMG do not address the problems of variable and potentially scarce bandwidth. Thus the Comm Server has been defined as a specialization of the more generic CORBA-type ORB. The Comm Server helps applications determine how much and what type of information they want to communicate and then enables them to exploit underlying ORB services as appropriate.

The mission of scientists is to suggest solutions that, if implemented by politicians, can improve international stability. This paper presents an object oriented model of the system for EW&RR in international context. Considered are four levels of the system, all comprising decision making and execution functions (Shalamanov, 1997). Three other levels are revealed: level of technical architecture (TA) of the system; level of the information management system (IMS) for document based group work; and level of the decision support system (DSS), where actual "business objects" are processed. These three levels are strongly connected with the levels of the cognitive process (observation => data => information => knowledge => understanding => decision => understanding => knowledge => information => data => action) of Command and Control (C2) and its simplified Observation-Orientation-Decision-Action cycle representation on Fig. 1.

The technical architecture level is related to data, observation and action. The information management system relates to information and orientation. The decision support system is connected with knowledge, understanding, and decision. There is no solid border between DSS and decision makers (DM)/staff. DSS and DM share common knowledge, but only DM can understand the situation and make responsible decisions. Of course, DM can delegate authority for some "simple decisions" to lower levels. Every one of these levels has its own set of objects. Every object can be one of the three types: object-entity (O-E), object-connection (O-C), and object-rule (O-R) (Shalamanov, 1992).

There are many governmental, i.e. DARPA/DISA Advanced Information Technology Services - Joint Program Office, and commercial, i.e., CORBA, initiatives in the field of building environment for object oriented representation of main information services and business processes that use these services. When the framework for the suggested three levels of OOM of EW&RR in international context is considered, strong reliance on opportunities to use achievements of the above programs is assumed.

The first two level objects are universal. Therefore, the analysis herein is focused on objects of the third level and the technology to use objects of second (MIS) and third (DSS) level. Among the early warning and rapid reaction specific third level *objects* are the following:

A. EW *objects*: *conflicts* and *indicators*; *collection measures* in *space* and *time*; *actors* to perform the above measures; *planning* techniques to develop *plan* for collection; *mapping* techniques to predict eventual conflict according to collected indicators values;

B. RR *objects*: *action measures* to prevent/resolve predicted/ongoing conflict; *planning* techniques to develop *plan* to synthesize adequate actors from

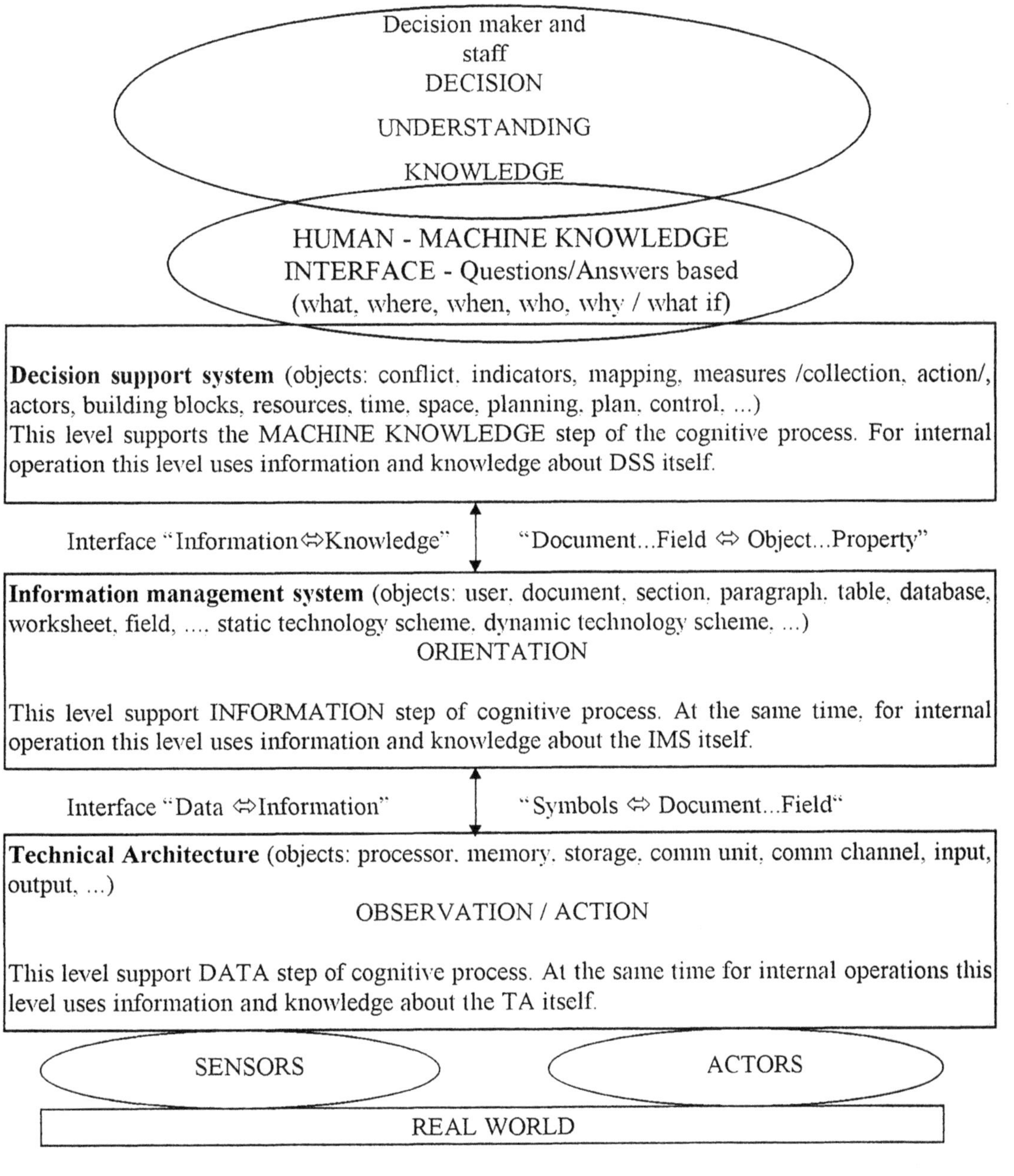

Fig. 1. Object-oriented model of the EW&RR with its three levels - TA, IMS, DSS, connected with the levels of the cognitive process.

- *building blocks* and *resources* to perform the above mentioned *action measures* in *space* and *time*; *planning* techniques to develop *plan* for operation as a sequence of *actions*; *control* techniques to guide the performance of the *plan* by *actors*.

Examples of second level specific *objects* are the Microsoft Office Objects and can be defined as *document* (standard and linked), *communications*, *procedure* (manual, automated, automatic). First level objects are sensor, computer, communication channel, disk, memory, etc. (Shalamanov, 1992). The aim is to develop an environment based on first level *objects* in the computer network that provide support for second level objects, connected with specification and manipulation of EW&RR specific *objects*. That means to be able easily and quickly to transform data from sensors in understanding trough building adequate third level model, to make decisions and to transform knowledge about the decision to data (commands) for actors to execute it.

All *objects* have their own *properties*, *methods*, associated *events/time* and *event/time-driven procedures*. In this environment, the staff of different specialists will perform its duties to support EW, consultations, situation analysis, collection planning and control, conflict prediction, force planing, operations planning and control, and other principal staff functions. An associated problem is connected with definition of staff organization, procedures, training and assessment.

The following discussion is focused on brief analysis of the EW problem, the RR problem, the presentation of the system for EW&RR as a network of situation centers and definition of a project for evolutionary system development through rapid prototyping.

3. CONFLICT EARLY WARNING PROBLEM

The essence of the conflict early warning (EW) problem is to find out a mapping between some set of indicators, values of which can be estimated, and the possibility for outbreak of a conflict. There are at least three issues connected with this problem: selection of indicators, categorization of conflicts, and mapping techniques. The analysis of these issues leads to further subdivision. For example, what kinds of conflict are considered (Brecke, 1997), how to choose indicators (Ivanova, *et.al.*, 1997; Borden, 1996), how to measure them, what kind of mapping techniques to use, what are the quantitative and qualitative characteristics of the mapping (Borden, 1996; Ivanova, *et.al.*, 1997; Tagarev, *et.al.*, 1996). Combining the solutions of these problems to build a consistent methodology for early warning is another problem of its own significance (Mueller and Law, 1995; Tagarev, *et.al.*, 1996).

On the other hand, the issues can be considered as scientific, technological and systemic. Scientific issues of the EW problem connected with the development of a taxonomy of conflicts are on the attention of researchers for long time, but the new security environment after Cold War adds new dimensions further complicating the investigations. Nevertheless, there are significant theoretical developments to support selection of indicators and to allow implementation in real systems (Borden, 1996; Brecke, 1997; Ivanova, *et.al.*, 1997).

The methodology of EW is based on different approaches to the mapping issue. One approach is based on integration of developments in neural, fuzzy and evolutionary methods, as well as methods from chaos theory for the purpose of EW of violent conflicts (Tagarev, *et.al.*, 1996). The authors study combinations of the learning capabilities of Artificial Neural Networks (ANN) with representational power of fuzzy sets theory to specify expert knowledge on a security situation. An ANN recognition scheme, "trained" on a historical data base is used to map indicator sets into conflict classes. Fuzzy set tools are used for interpretation of the ANN outputs. Methods from chaos theory and evolutionary methods are used to determine a representative set of input indicators and to optimize the ANN's topology, its parameters, as well as the parameters of the learning algorithm.

Regarding the technological problem, there are sufficient collection, storage, communications, computational and presentation power to implement the above mentioned methodology, as well system integration capabilities to design and implement operable systems. The efforts are currently concentrated on the development of methodology and tools for conflict specification, indicator selection, and mapping. The integration of efforts in this field will permit to solve diverse EW issues.

4. RAPID REACTION - CJTF CONCEPT

The reaction for conflict prevention/resolution can be multidimensional, but as a rule it comprises a military element. The concept of Combined Joint Task Force (CJTF) was developed in the early 90s as a tool for military reaction to the new security challenges. The concept undergoes further development both in NATO/PfP exercises and real-life implementation in the IFOR/SFOR operations.

CJTF for MOOTW are NATO-led, but many non-NATO nations (mainly PfP partners) also participate. Effective use of components from these countries may be provided in result of complex certification, EW, consultation and force&operation planning processes. One of the most important problem in the implementation of these forces, including preparation, deployment, employment, rotation, redeployment, is their integrated deployable C4I system, connected with the international system for EW, political consultations and force planning.

The RR model consists of seventeen principal functional areas of concern for the CJTF (Hayes-Roth and Erman, 1994): Stand-up (and -down) the CJTF staff and systems; Receive orders and clarify mission; Send reports; Set up and operate communication links; Organize activities, assign tasks, and manage CJTF staff; Set objectives and constraints; Assess the situation; Generate and evaluate alternative plans; Package and communicate assessments and plans; Request information; Supply information; Evaluate and combine pieces (for assessments, plans, and communications); Give orders and clarifications; Accept reports; Monitor for and adapt to significant changes; Use and update (computerized) data and objects; Run computer models.
Although each functional area interacts with most of the others, the seventeen areas are highly modular with respect to relevant concerns and responsibilities, which can help to divide and conquer the domain modeling problem. Examining the CJTF concept as a tool for rapid reaction and striving to shape the information environment to support the process, there are some well defined issues that require solutions:

- capability based force planning, consistent with the NATO Planning and Review Process (Davis, 1995);
- operations planning on the basis of early warning and using certified building blocks, consistent with the NATO operations planing procedures;
- certification of building blocks and consultation during force and operations planning, consistent with NATO standards (STANAG) and the evolving NATO/ EAPC consultation system;
- C4I support for deployable CJTF to execute MOOTW, that leads to interoperable C2 system.

Again, there are scientific, technological and systemic issues. The STC Operations Research Division science support activity is a typical example of how to cope with the force and operations planing problems both scientifically and technologically. Similar experience is available on national level for every NATO/PfP country. There are many data bases and decision support tools for registration and manipulation of force and operations related information, as well as software packages to facilitate decision making during the planning process. Standardization of documents, procedures, force building blocks, resources and other elements of forces and planning/control process is in progress, too.

The current efforts are concentrated on the main problem of planning (both force and operations). The planning methodology is based on tools for:

- definition of goal (mission), tasks (distributed in space and time), mapping the tasks with real objects in space and time, as well as standard actions to be executed on the objects;
- assessment of the force capabilities to execute the actions; synthesis of main and support force elements;
- global distribution of force elements to the objects;
- secondary distribution of the force main elements (collectors/actors) to tasks, zones, time intervals according to the global distribution (previously defined) and subject to additional constraints for implementation of collectors/actors;
- distribution of support elements according to evolving tasks to support the main elements;
- synthesis of the C4I system, adequate to the distribution of main and support force elements to tasks, zones and objects (actions).

Main methods are information (data base and expert systems) methods and operations research (dialog iterative discrete optimization) methods as well as "What if" analysis and simulation. The opportunity for cooperative development of this methodology is also significant.

5. SIMULATION EXPERIENCE

The efforts to implement the model under consideration are divided in the following steps:

- Developing open office system - Management Information System (MIS) based on Windows NT (Back Office Servers) and MS Office to investigate the implementation of first and second level commercial standards for development of EW&RR system;
- Shaping the MIS, using formalized documents, language to specify information-functional (static technological schemes) and time-event (dynamic technological schemes) logic of EW&RR activity and development of intranet, HTML based environment of operational documents - building of a powerful second level;
- Integrating the help/advise hypermedia environment of doctrines, manuals, and previous operational documents to support DM.

These steps are based on experience and documents of many NATO/PfP exercises for crisis management and conducting MOOTW (Shalamanov, 1997). Next step is to implement previous experience in hybrid heterogeneous expert systems (Shalamanov, 1992) to build powerful third level of OOM for EW&RR system.

Experiments were performed to predict development of interstate conflict using about 250 copies of 30 object-classes (O-E and O-C) and about 50 object-rules, using CLIPS language (computer language for intelligent planning systems - forward reasoning).

Another type of experience was acquired using soft-computing and tree based inference to predict (classify) the state of certain objects' in EW context. Both approaches are based on analysis and processing of large amount of examples that connect values of a set of indicators to certain state of the considered object. Further studies require integration of the three levels and more "business" expertise.

6. POTENTIAL DETERRING EFFECTS

The final goal is to create a distributed multinational system that supports EW, consultations, force and operations planning, operational command and control of CJTF for MOOTW. Such a system is typical for coalition consultation, command and control system which has to take into account external (physical environment and situation, political and policy, socioeconomic, and technological), human, internal (planning and execution of operations), interoperability factors (Maurer, 1994).

There are achievements and practical results in different aspect of EW&RR. Situation Centers (under different names) with similar functions exist in all NATO/PfP countries. Solution of the complex problems may be found in integration of science, methodology, technology, organizational, financial and political resources in one system - a *Network of Situation Centers* (SC) for EW&RR control, augmented with a flexible, deployable C4I system for CJTF.

On national level, the SC integrates national efforts in EW and force planning. These SCs collect information about conflicts' indicators values and process them to "produce" EW. At the same time, the SCs collect information about the status of certified force building blocks and resources/facilities of the country and support force and operations planing.

The National SC is a tool for international consultations in the international network of SCs. The network aids international EW, collective decision-making for RR and consequent CJTF force planning and MOOTW operation planning, as well as planning and control of deployment, implementation, and redeployment. Shalamanov (1997) provides detailed description of the system of NSC and the problems related to its development.

Summarized, the main deterrent effects of EW&RR system for potential belligerent are:

- "The international community looks at me and knows all my actions and, probably, intentions";
- "It can react quickly and will react, because, maintaining such a system, it is committed to peace";
- With a sufficiently powerful DSS (as a result of acquired knowledge, information received in real-time and learning capabilities) the international community may find solutions not only for fighting occurring threats but also for preventing threats.

7. CONCLUSION

The development of EW&RR planning and execution control system may contribute to improving international stability. The object oriented model of the system reveals three levels: technical architecture, MIS, and DSS of the specific activity of EW&RR and complex interaction of scientific, technological and systemic factors.

Enhanced PfP/EAPC efforts can play a crucial role in the search for solutions by supporting the development of a network of situation centers and the use of the CJTF concept not only for NATO/WEU adaptation, but as a framework of partner participation in the military pillar of the European security system.

Practical steps would be taken after official suggestion from interested and motivated organizations. Hopefully, such steps will be taken by UN or NATO authorities in the framework of EAPC/PfP and they will help to consolidate national efforts in support of the project of developing international Network of Situation Centers.

Acknowledgment. The work on this paper was partially supported by Bulgarian National Science Fund under grant I-621-96.

REFERENCES

Borden, A. (1996). An Information Warfare Roadmap. In: *Information Aspects of Security and Development of Modern Societies*, pp. 5-16. AFCEA-Sofia, Sofia.

Brecke, P. (1997). Defining the Dependent Variable for Conflict Early Warning: A Taxonomy of Violent Conflicts, Annual Meeting of the International Studies Association, Toronto, March 18-22, 1997.

Davis, P. (1995). *Institutionalizing Planning for Adaptiveness, New Challenges for Defense Planning*, RAND, pp. 73-100.

Erdem, H. and J. Hellvik (1993). Combined Joint Task Force (CJTF) - Communications and Information Systems Issues and Solutions for a Land-Based Scenario. In: *Peacekeeping - a Partnership for Peace Challenge*, AFCEA Europe Symposium, 19-21 Oct 1993, pp. 46-51.

Hayes-Roth, F. and L. Erman (1994). *The Joint Task Force Architecture Specification (JTFAS)*. Teknowledge Federal Systems, Palo Alto, CA.

Ivanova P., T. Tagarev and A. Hunter (1997). Selection of Indicators for Early Warning of Violent Political Conflicts by Genetic Algorithms. *TRANSITION'97*, Warsaw, 1997.

Kile, F. (1996). Dynamic Peace as Goal. *IFAC-SWIIS NEWSLETTER*, Vienna, October 1996.

Maurer, M. (1994). *Coalition Command and Control*, National Defense University, Forth McNair, Washington, DC.

Mueller, A. and D. Law (1995). Crisis Management and Conflict Prevention in a Historic Transition Period. *UNIDIR News Letter*, **30**, pp. 27-38.

Shalamanov, V. (1992). Environment and Methodology for Development of Intelligent Systems for Complex Situation Assessment and Prediction, In: *AFCEA Europe Prague Seminar*, September 1992, pp. 31-34.

Shalamanov, V. (1997). CJTF C4I System for Early Warning and Rapid Reaction. *18th AFCEA Europe Brussels Symposium*, October 1997.

Tagarev T., P. Ivanova and A. Moscardini (1996). Soft-Computing Methods for Early Warning of Violent Conflicts. In: *Information Aspects of Security and Development of Modern Societies*, pp. 44-51. AFCEA Sofia, Sofia.

A BEHAVIOR MODEL FOR PREVENTIVE HEALTH PRACTICES

Gabriela Simion

Departament of Automatic Control and Engineering Systems
Univesity Polytechnica of Bucharest
e- mail: simion@freddy.cib.pub.ro

Abstract: In our everyday social live we routinely understand the behaviour of others by atributing mental states to them. The significance of behavioural model and life style for health is also now widely acknowledged. The paper presents firsts the concept of life style, its components and its relevance for the major consense of professional diseases . The presented model is important for widespread education on preventive health practice. *Copyright © 1998 IFAC*

Keywords:behavior, knowledge representation, cognitive systems,data processing, matrix, model approximation,decision making.

1. INTRODUCTION

Nowadays, importance of behavior factors in preventive health has increased due to uniforseeable character of events and individual psychosocial solicitations. Nature of diseases significantly changed, because of modern medical knowledges, but also of fast social modifications as industrialisation, informational impact, townlife. So heart diseases, cancer,diabetis, aids are civilisation diseases.

Identification and dynamics of risk factors for these diseases lead to necessity of a multidisciplinary work and both psycho-behavioural variables and traditional are considered as important factors for disease evolution. These considerations point to realise a cognitive behavior model. Such a model is designed to describe a large range of beliefs concerning the importance of behavior and life style for health.

Life style is defined (Dean, 1989) as sume of decisions and voluntary acts which affects health state and there are two types of life style : negative and positive. Risk factors for diseases are behavior variables which are characteristic for negative lyfe style.

2.BEHAVIOR MODEL

A behavior model is described by aiding cognitive factors (beliefs and knowledges) and reason factors. Adoption of a healthy life style need analysis of model composed of three stages: information processing , inference (reason stage) and implementation of decisions.

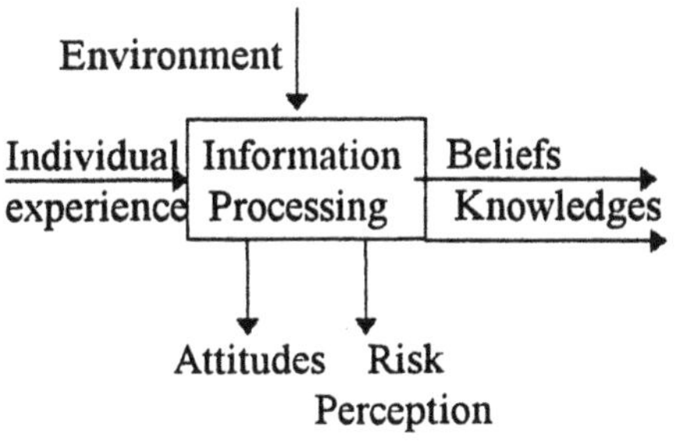

Fig. 1 First stage of model

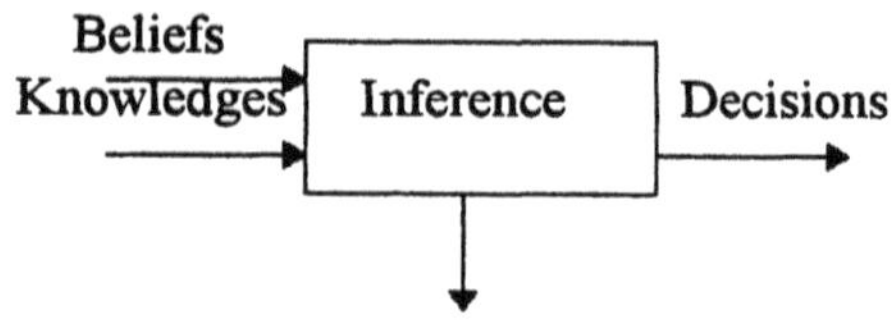

Fig. 2 Second stage of model

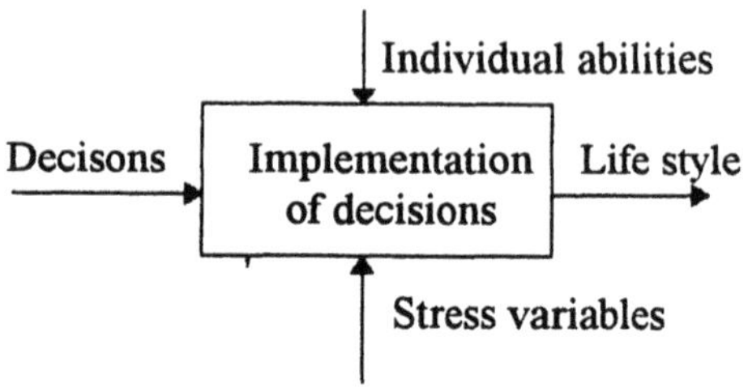

Fig. 3 Third stage of model

Classic feature model is a fruitful begining point for designing a cognitive system revealing importance of lyfe style for health care preventing. The way for modelling verbal knowledges is very important . An extension of features model is realised , because both analysis are made , for features (behavior variables) but also for relations between determined categories (diseases).

First step in designing of model is to collect a suitable list of behavior variables relevant to disease states: overweight, stress, animal fat consumptions, coffee consumption, heredity, smoking, alcohol drinking, phisical exercises, sunlight expose, drugs. A series of statements about diseases (heart disease, arterial hipertension, phisical disease, cancer) made by informants are collected from ethnographic interwies. Informants use a multiple point scale with different graduates : definetly, probably, possibly, probably not, definitely not. Ratings are used in attempt to measure individual differences in personality. A five scale of behavior factors rating (Mamdani and Efstathiou, 1984) is represented as below:

definetly not				definetly
-2	-1	0	1	2

Fig. 4 A five scale of behavior factors rating

To simplify data the reponses are dichotomized by using the mean score for each belief about importance of a behaviour variable for diseases and it is decided whether a score is considered a "yes" or "no". Item by attribute matrix **M** is obtained. Attributes are behavior factors and items are diseases. Each column of this matrix has element 1 for a "yes" response concerning belief that a behavior is important for producing a disease and 0 is for a "no" answer.

Table 1 Matrix of item by attribute for diseases and behavior vaiables

	OVERWEIGHT	STRESS	ANIMALFAT	HEREDITY	TOBACCO	ALCOHOL	PHISICALEX	SUNLIGHTEXP
HEART DISEASE	1	1	1	1	1	1	1	0
HYPERTENSION	1	1	1	0	1	1	0	0
PSIHICAL DISEASE	0	1	0	1	1	1	0	0
LUNG CANCER	0	0	0	1	1	0	0	0
SKIN CANCER	1	0	0	1	0	0	0	1

It is considered (Băban ,1997) a sample in study formed of N= 394 persons and probabilities of behavior risk variables are determined.

Table 2. Beliefs concerning risk behavior factors for health

Diseases	Behavior (risk) variable	P %
Heart Disease	Overweight	81.9
	Stress	80.8
	Animal fat consumption	68.8
	Heredity	64.4
	Tobacco	62.3
	Alcohol	55.1
	Phisical exercise	10.6
Arterial Hypertens.	Overweight	65.5
	Stress	87.9
	Tobacco	48.0
	Animal fat consumption	65.5
Psihical diseases	Heredity	71.5
	Stress	87.9
	Alcohol	61.7
Lung cancer	Tobacco	97.8

	Heredity	27.3
Skin cancer	Sunlight exposure	15.3
	Heredity	43.8

Behavior variables with probabilities less 10% are not considered.

Matrix **M** is used for determination of the best behavior decision , by aiding multiattribute stategy.

A set of behavior decisions **V** and a set of criteria **C** is considered:

$$\mathbf{V} = (V_1, V_2,..., V_m) \quad (1)$$

$$\mathbf{C} = (C_1 , C_2 ,..., C_n) \quad (2)$$

Consequences matrix is **A** , with a_{ij} elements representing evaluation of v_i decision for c_j criterion:

$$\begin{pmatrix} a_{11} & a_{12} & a_{1n} \\ a_{m1} & & a_{mn} \end{pmatrix} \quad (2)$$

Weights of criteria p_{0j} and d_j importance oefficients, are evaluated (Andrasiu, et al.,1986) by aiding entropy method :

$$hj = - \frac{1}{\ln (m)} \sum_{i=1}^{m} r_{ij} \ln (r_{ij}) \quad (3)$$

for j=1 to n and

$$r_{ij} = \frac{a_{ij}}{\sqrt{\sum_{i=1}^{m} a^2{}_{ij}}} \quad (4)$$

$$d_j = 1 - h_j \quad (5)$$

$$p_j = \frac{d_j}{\sum_{j=1}^{n} d_j} \quad (6)$$

This computing is related on beliefs probabilities for studied people and their subjectivy. Beliefs about importance of behaviors for health were associated with the occurrence of behavior. Ierarhisation of decisions is realised with an alghorithm based on aditive weights method. λ_j are subjective importance coefficients. Weights for aditive method are computed with following algorithm:

for j=1 to n

$$p_{0j} = \frac{\lambda_j p_j}{\sum_{j=1}^{n} \lambda_j p_j} \quad (7)$$

Utility function for each decision is determined :

$$U_i = \sum_{j=1}^{n} p_{0j} r_{ij} \quad (8)$$

Decision V_k is the best if condition (9) is verified :

$$U_k = \max U_i \text{ for } i = 1 \text{ to } m \quad (9)$$

3. CONCLUSIONS

The aim of this paper is to design a model for representation of beliefs concerning implication of behavior variables for health. The development of item by attribute matrix includes that it is more than just a representation , it is also a processor of a special kind. From this perspective, model is a kind of mental inference representation , which creates a complex interpretation from minimal inputs.

REFERENCES

Andrasiu, M., Baciu, A. ,Pascu, A.,Puscas, E.and Tasnadi,Al. (1986). *Metode de decizii multi-criteriale* (Editura tehnică), Bucuresti.

Băban,C. (1997). Comportament si sănătate . *Creier.Cognitie .Memorie*, **3**, 37-52.

Dean, K. (1989). Self-care components of lyfestyles: the importance of gender, attitudes and the social situation. *Social Science Medicine*, **29**, 137-152

Mamdani, E.H., Efstathiou, J. (1984). An Analysis of formal logics as inference mechanisms in expert systems *Int. J. Man-Machine Studies*, **21,** 213-227.

ACTS OF MATHEMATICS WITH HIERARCHICAL HAZE

Svetlana Novikava, Andrey Zhybul

Hierarchical Multilevel Systems Laboratory,
Institute of Mathematics&Cybernetics,
Minsk, Belarus
220064, P.o. box 48, e-mail: imc@org.by, hmsl@buk.lingvo.minsk.by

Abstract: The paper considers mathematical acts (uniting&multiplying) with hierarchical time&space. All its strata (times, levels) contain their own haze (chaos, uncertainty, vagueness). The new mechanism of mathematics allows to define the process when chaotic fields of diverse strata turn into well ordered things of higher strata with new haze in their interactions and with wider horizons of their might. Suggested mechanism includes habitual means which work with a chaos, and it is more cohered with real tasks in physical, chemical, biological demographical, engineering, knowledge, and thought (arising) strata. Its main abilities are accounted in the paper by the instances of active and waning systems in hierarchical time&space.

Keywords: hierarchical mathematics, haze, active systems, waning reasons

1. NEW MECHANISM OF MATHEMATICS AND ITS ACTS WITH HIERARCHICAL HAZE

Now the main law of existent hierarchical time&space is described by the following statement: all strata arise in lower strata multiplying and uniting , create more high strata and are changed by higher strata activity. This statement allows to define exact statute A^{λ} of hierarchical mathematics - aed theory which will be changed by the mechanisms of new strata constructed with it (Novikava, et al, 1997).

Mechanism of sway (might) realizing in hierarchical time&space connects the main directions of cybernetics. They are design, control and learning - the stages of sway process. Design stage answers to uniting act and learning one answers to multiplying act (control process connects them). These acts are realized in existent haze (vagueness, chaos) and create the new one — new strata of hierarchical time&spave always belong to hazy areas with high hazard. The mechanism of new mathematics (theory of sway) must decrease a danger of errors in haze.

Aed statute A^{λ} has two its main hierarchical symbol images ${}^{\times}\alpha^{\lambda}$ and ${}^{+}\alpha^{\lambda}$ which are cohered with the acts of multiplying and uniting in aed mathematics. They connect the own strata (directions, details) of A^{λ}. Aed strata are: Λ,λ-level (hierarchical number, time&space), Γ,γ-statute (law, connection), Ρ,ρ- act (process), Ω, ω - unit (state) Σ,σ - construction (contents), Β, β - hierarchical haze (new time, arising level), Α, α - sway (coordinator).

Aed statute A^{λ} in time l is described by its symbol image ${}^{\times}\alpha^{\lambda}$ in following way:
${}^{\times}\alpha^{\lambda}$:

$$A^{\lambda} \xleftrightarrow[\rho]{\gamma} \left\{ {}^{\beta\;\gamma}_{\omega}\underset{\rho}{A}{}^{\lambda}_{\sigma} \right\}^{\lambda} \xrightarrow[\rho]{A^{\lambda}} \beta$$

$$\Lambda^{\lambda} \overset{\gamma}{\underset{\rho}{\longleftrightarrow}} \left\{ {}^{\beta}_{\omega} \overset{\gamma}{\underset{\rho}{\Lambda}} {}^{\lambda}_{\sigma} \right\}^{\lambda \xrightarrow[\rho]{\Lambda^{\lambda}} \beta}$$

$$\Gamma^{\lambda} \overset{\gamma}{\underset{\rho}{\longleftrightarrow}} \left\{ {}^{\beta}_{\omega} \overset{\gamma}{\underset{\rho}{\Gamma}} {}^{\lambda}_{\sigma} \right\}^{\lambda \xrightarrow[\rho]{\Gamma^{\lambda}} \beta}$$

$$P^{\lambda} \overset{\gamma}{\underset{\rho}{\longleftrightarrow}} \left\{ {}^{\beta}_{\omega} \overset{\gamma}{\underset{\rho}{P}} {}^{\lambda}_{\sigma} \right\}^{\lambda \xrightarrow[\rho]{P^{\lambda}} \beta}$$

$$\Omega^{\lambda} \overset{\gamma}{\underset{\rho}{\longleftrightarrow}} \left\{ {}^{\beta}_{\omega} \overset{\gamma}{\underset{\rho}{\Omega}} {}^{\lambda}_{\sigma} \right\}^{\lambda \xrightarrow[\rho]{\Omega^{\lambda}} \beta}$$

$$\Sigma^{\lambda} \overset{\gamma}{\underset{\rho}{\longleftrightarrow}} \left\{ {}^{\beta}_{\omega} \overset{\gamma}{\underset{\rho}{\Sigma}} {}^{\lambda}_{\sigma} \right\}^{\lambda \xrightarrow[\rho]{\Sigma^{\lambda}} \beta}$$

$$B^{\lambda} \overset{\gamma}{\underset{\rho}{\longleftrightarrow}} \left\{ {}^{\beta}_{\omega} \overset{\gamma}{\underset{\rho}{B}} {}^{\lambda}_{\sigma} \right\}^{\lambda \xrightarrow[\rho]{B^{\lambda}} \beta}$$

$$A^{\beta} \overset{\gamma}{\underset{\rho}{\longleftrightarrow}} \left\{ {}^{?}_{\omega} \overset{\gamma}{\underset{\rho}{A}} {}^{\beta}_{\sigma} \right\}^{\beta \xrightarrow[\rho]{A^{\beta}} ?}$$

In this way all aed strata can renovate its original unit A^{λ}, they have all its signs and abilities. At the same time they have individual lines, that is their space is the area of diversity of A^{λ} states. The strata $\Lambda, \Gamma, B, P, \Sigma, \Omega$ (constructed by the multiplying of A^{λ} in the details $\alpha,\beta,\gamma,\rho,\sigma,\omega$ of its construction) are connected both by their original unit A^{λ} and by the details of their own constructions (by their new interactions). Thanks to that all aed strata may be renovated and amended when any stratum is changed.

The acts ${}^{\times}\rho^{\lambda}$ and ${}^{+}\rho^{\lambda}$ of original (key) unit A^{λ} multiplying and then new ordinary units ω_{τ}^{λ} uniting in σ^{λ} with connections γ^{λ} lead to the arising of the key unit of new time A^{β} which is defined in stratum β^{λ} of hierarchical haze. The unit A^{β} contains dark haze in its image (it is signed by the symbol ?, and it will be defined in time β).

The image ${}^{+}\alpha^{\lambda}$ (Fig.1, scheme of aed processor) allows to see the acts of multiplying&uniting, their bases, aims and connections. In Fig 1: A^{λ} - original state of aed ($A^{\lambda} \leftrightarrow \omega^{\lambda}$), A^{β} - its leading state in time λ; act ${}^{\times}\rho^{\lambda}$ multiplies the original order γ^{λ} (original statute (code) of A^{λ}) in the units $\omega^{\lambda_{\tau}}$ of construction $\sigma^{\lambda_{\tau}}$ created in time λ_{τ}; thanks to that the lower strata units $\omega^{\lambda_{\tau}}$ turn into the ordinary units ω_{τ}^{λ} in the new construction σ^{λ}; act ${}^{+}\rho^{\lambda}$ in σ^{λ} unites the ordinary units ω_{τ}^{λ}, and constructs the new sway $\omega^{\beta} \leftrightarrow A^{\beta}$; in the process ρ^{λ} the image of hazy time β^{λ} has two strata - ${}^{\times}\beta^{\lambda}$ and ${}^{+}\beta^{\lambda}$ which answer two stages of this process: ${}^{\times}\rho^{\lambda}$ and ${}^{+}\rho^{\lambda}$.

The images (codes, records) ${}^{\times}\alpha^{\lambda}$ and ${}^{+}\alpha^{\lambda}$ are convenient both in the work with mathematical theories, and in the defining of concrete world strata (physical, chemical, biological, demographical engineering, knowledge, and thought (new) times).

Graphical scheme of Fig.2 is other way of accounting of time increasing process when aed strata are turned one into another and hazy area is changed. Time λ (time of sway of stratum λ) contains two times: ${}^{\times}\lambda$ (time of multiplying) and ${}^{+}\lambda$ (time of uniting). It connects the earlier ($\lambda_{\tau} \leftrightarrow \tilde{\lambda}$)and new (β) times, where λ_{τ} — times of λ stratum arising, β — times of λ stratum renovating and changing.

The area of haze in hierarchical images ${}^{\times}\alpha^{\lambda}$ and ${}^{+}\alpha^{\lambda}$ is β^{λ}. The stratum β^{λ} (aims, arising times) is changed in time incresing process. The stages of its changing are defined by Fig.3 which allows to see the contents of hierarchical haze and mechanisms $\rho^{\tilde{\lambda}}$ -, ρ^{λ} of work with it. The mechanism $\rho^{\tilde{\lambda}}$ is technology of lower times (when chimerical statute $\gamma_{\sigma}^{\tilde{\lambda}}$ of unit $\omega_{0}^{\lambda} \leftrightarrow A^{\lambda}$ and then this unit are constructed); while ρ^{λ} multiplies original unit ω_{0}^{λ} (creates $\sigma^{\lambda} \leftrightarrow \left\{\left\{\omega_{i}^{\lambda}\right\}, \gamma_{\sigma}^{\lambda}\right\}$), and unites ω_{i}^{λ} (constructs $\omega^{\beta} \leftrightarrow A^{\beta}$, discovers time β). The mechanism ρ^{λ} which is realized in vague ranges belongs to art strategies). Fig.3 connects the stages of hierarchical haze changing. Its states ($\beta^{\tilde{\lambda}}, \beta^{\lambda}, \beta^{\beta}$) in diverse times ($\tilde{\lambda}, \lambda, \beta$) have diverse contents - the new horizons in these times.

The new sway strata of hierarchical time&space always arise in the area of haze and then they

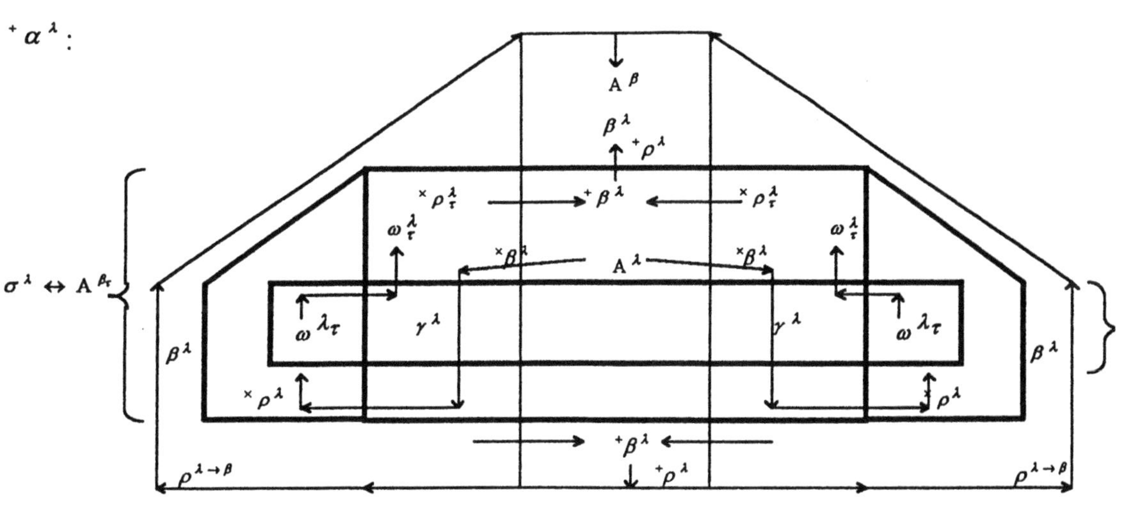

Fig.1 Graphical symbol-image of aed (the scheme of aed processor). Hierarchical haze β^{λ} is the area strata constructing

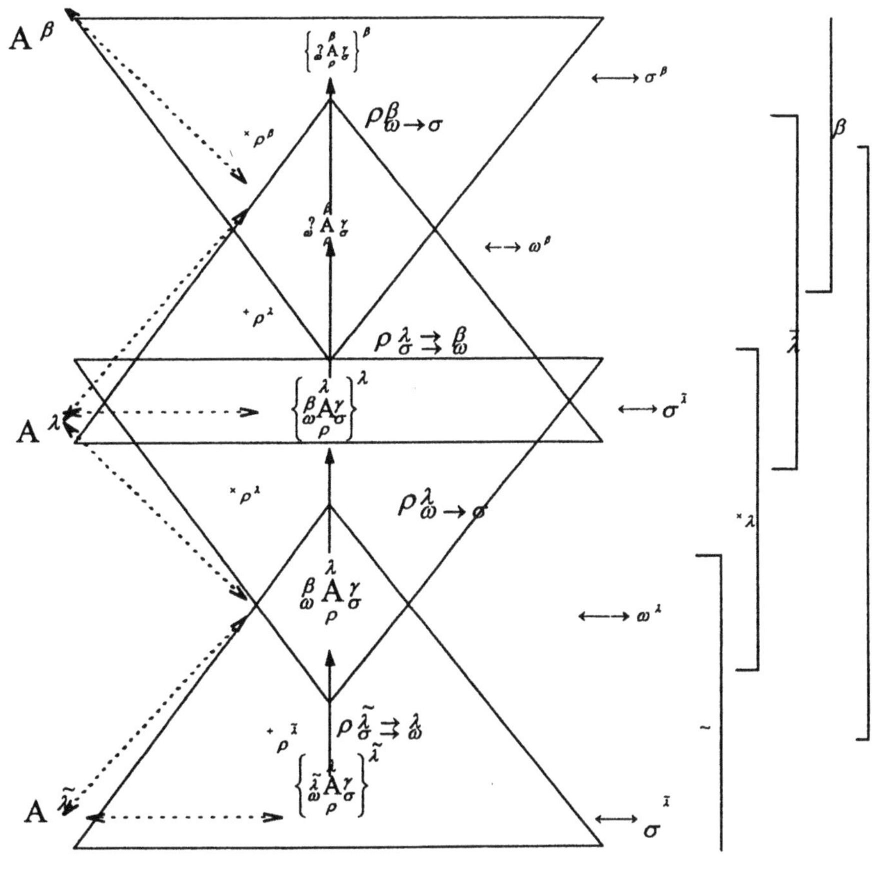

Fig 2. The connections of aed strata in the process of hierarchical time increasing; hierarchical haze is signed by symbols β and ?; the changes of hazy areas have to change the main images of aed mathematics when its original state A^{λ} turns into new one A^{β}.

2. SYSTEMS WITH WANING REASONS

In agreement with the main law of hierarchical time&space the right (worthy) sway systems in time increasing process must maintain (moreover they must construct) areas of haze whose work will change these systems. In its turn a hazy area (as active strata of sway) must construct the new things of higher times with well defined statutes and

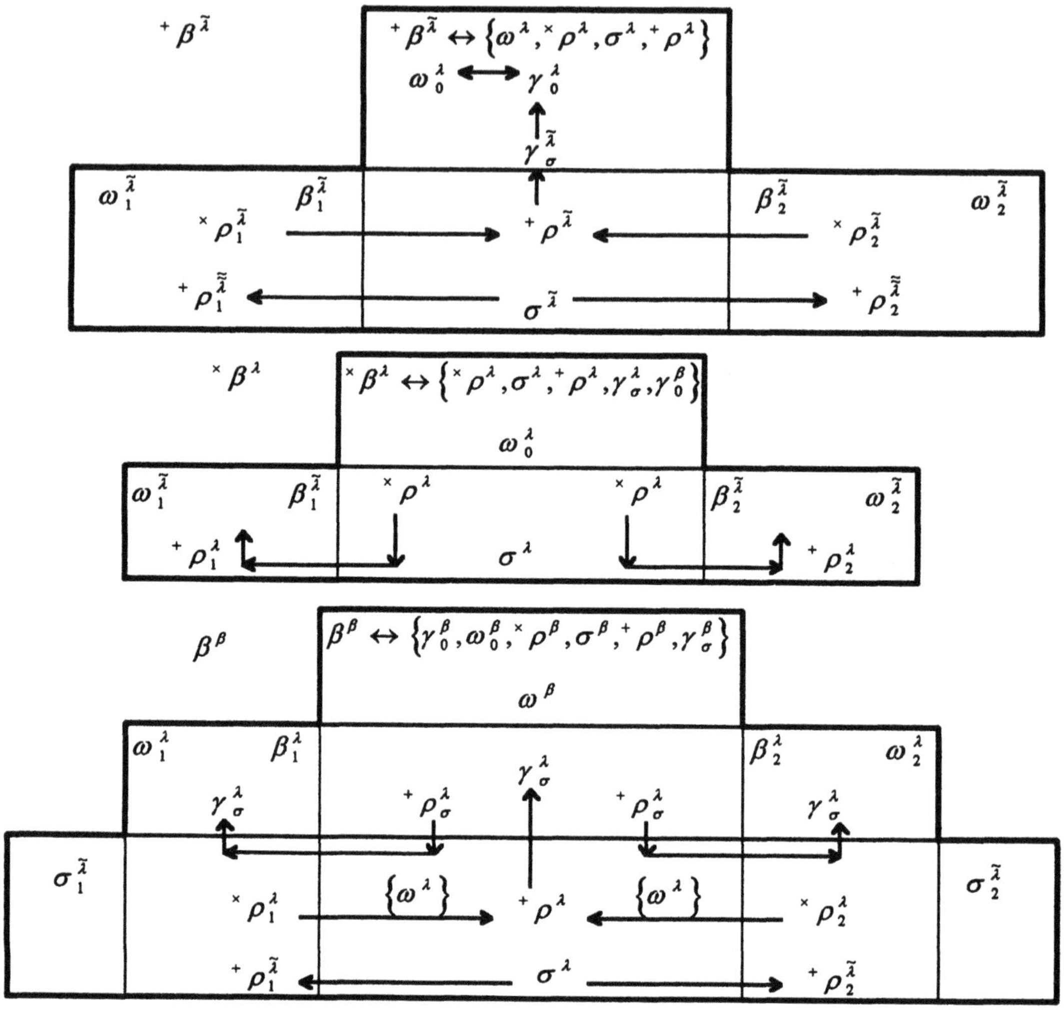

Fig. 3. Acts of uniting $({}^{+}\rho^{\tilde\lambda}, {}^{+}\rho^{\lambda})$ and multiplying $({}^{\times}\rho^{\tilde\lambda}, {}^{\times}\rho^{\lambda})$ in aed processor with hierarchical haze $(\beta^{\tilde\lambda}, \beta^{\lambda}, \beta^{\beta})$; the contents of β^{λ} are changed in this process; uniting ${}^{+}\rho^{\lambda}$ and multiplying ${}^{\times}\rho^{\lambda}$ in time λ are hesitant acts of design&learning (creative work, art); at that time the acts of earlier time $\rho^{\tilde\lambda}$ belong to well defined technologies which are renovated by the sway ω^{λ} of time λ; its statutes $\gamma^{\lambda}_{\omega}$ change the laws $\gamma^{\tilde\lambda}$ of earlier times $\tilde\lambda$; however the new construction σ^{λ} contains the archaic strata $\sigma^{\tilde\lambda}$ whose laws are in force in their area; thanks to that the errors of the new stratum λ may be corrected when art activity leads to destroying process.

technologies of sway realizing in all existent strata.

However the main law admits the existing of waning systems which, unlike right ones, cannot amend their contents and activity. There are two main kinds of waning systems: monopolistic units ${}^{\Gamma}\omega^{\tilde\lambda}_{\tau}$ and odd constructions ${}^{\Gamma}\sigma^{\tilde\lambda}$. Their general characteristics are the errors in their areas of haze.

The units ${}^{\Gamma}\omega^{\tilde\lambda}_{\tau}$ which cannot construct the new hierarchical haze ${}_{\omega}\beta^{\tilde\lambda}$ have waning (limited) aims ${}^{\Gamma}_{\omega}\beta^{\tilde\lambda}_{\tau}$. They regard their own laws $\gamma^{\tilde\lambda}_{\tau}$ (may be the laws $\gamma^{\tilde\lambda_\tau}$ of their lower times $\tilde\lambda_{\tau}$) as the best in all times. That is they are aimed to the hind times. Their own images ${}^{\Gamma}\beta^{\tilde\lambda}_{\tau}$ of their future are waning, they contain old technologies and statutes ${}^{\Gamma}_{\omega}\beta^{\tilde\lambda}_{\tau} \leftrightarrow \{{}^{+}\rho, \gamma\}^{\tilde\lambda_\tau}_{\tau}$ where ${}^{+}\rho^{\tilde\lambda_\tau}_{\tau}$ and $\gamma^{\tilde\lambda_\tau}_{\tau}$ belong to hind times. Owing to that the systems

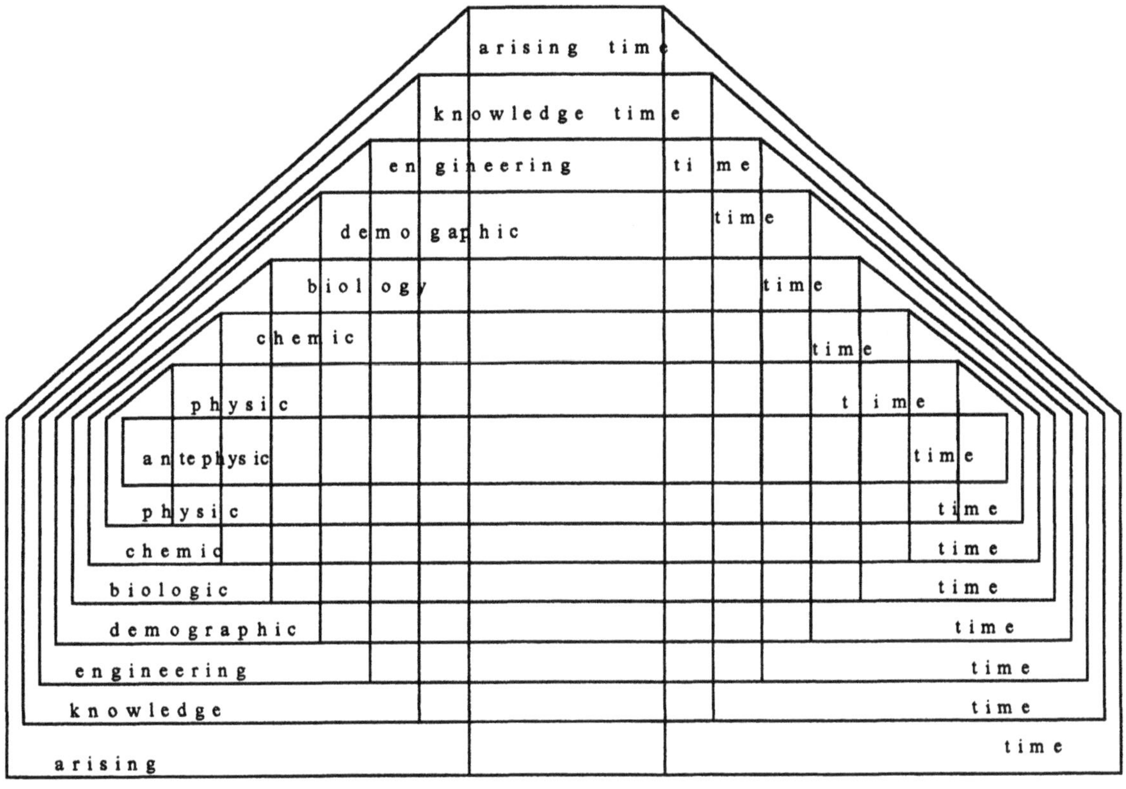

Fig.4 Aed statute in graphic number code: the image of hierarchic time&space.
The images of higher times arise in the areas of hierarchical haze of active (right) systems of all strata as single units with chimeric statutes (defined only by their own contents). Now the arising (thought) time unit is in this state. It is the field of interactions of knowledge systems (they are imaged as two units: to the right and to the left of new time). The new unit has direct contacts with all lower times (strata): its image intersects them and contains them in its sway field bounded by the inclined lines. The units interact by all their strata; their sway (higher) strata design&control their acts. The number of sway strata of all levels units is measured by the number of vertical lines in their constructions (by the distances of their times of sway from current sway); the number of their lower strata is measured by the number of horizontal lines (by the distances of their times of sway from the origin of times). Hazy area of thought (range of new sway arising) is zone of high hazard in activity of hierarchical systems.

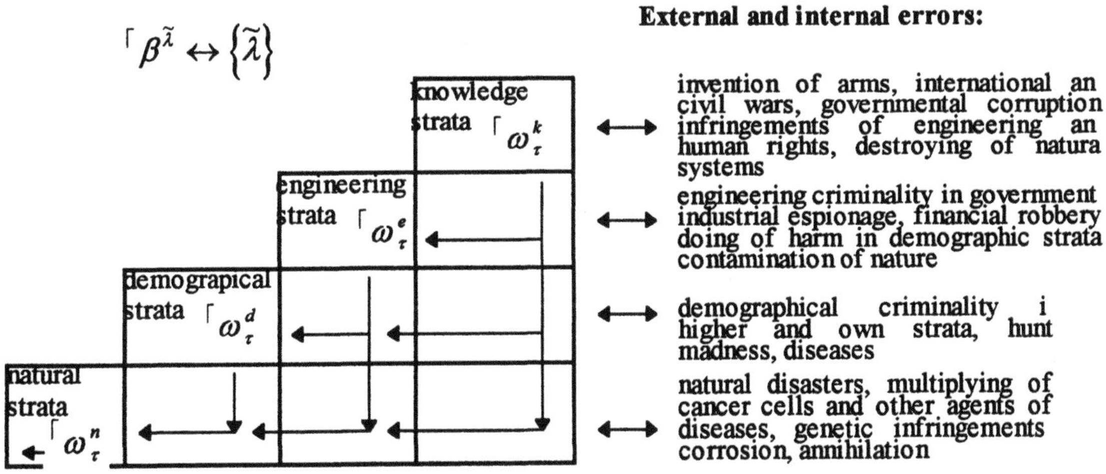

Fig.5. Waning systems and their interactions; the units $^{\ulcorner}\omega_{\tau}^{\tilde{\lambda}}$, $\tilde{\lambda} \leftrightarrow \{n, d, e, k\}$ have aims (hazy areas in their reasons) limited by their own and lower levels; they cannot construct the new strata and since that they are destroyed by their own activity within their constructions and/or in their contacts.

$^{\Gamma}\omega_{\tau}^{\tilde{\lambda}}$ with waning reasons $^{\Gamma}_{\omega}\beta_{\tau}^{\tilde{\lambda}}$ ban the multiplying of their own statutes (ban their changing). They are monopolistic units able to destroy an unaccountable (for them) things both in their environment and in their own contents. Even advanced means may be used by them as a weapon. They are destroyed in line

with the limits in their aims which hinder them to move to real sway strata in hierarchical time&space.

Odd constructions $^{\Gamma}\sigma^{\tilde{\lambda}}$ with lack of uniting ability cannot construct new sway ω^{λ} in their hazy areas $^{\Gamma}\beta_{\sigma}^{\tilde{\lambda}}$ which have not images of $^{+}\rho^{\tilde{\lambda}}$ and ω^{λ}: $^{\Gamma}\beta_{\sigma}^{\tilde{\lambda}} \leftrightarrow \left\{ ^{\times}\rho, \gamma \right\}^{\tilde{\lambda}}$ where $^{\times}\rho^{\tilde{\lambda}}$ multiplies the origin $\omega_{0}^{\tilde{\lambda}}$, but $\gamma^{\tilde{\lambda}}$ bans the uniting of $\omega_{\tau}^{\tilde{\lambda}}$, that is it maintains the old chaos. Owing to that the details $\omega_{\tau}^{\tilde{\lambda}}$ of odd systems $^{\Gamma}\sigma^{\tilde{\lambda}}$ begin to believe that the laws of earlier strata (the laws of their own arising) are stronger than ones in odd construction which cannot control and amend their interactions. Since that the aims of these details became limited, they incline to corruption (to higher aims replacing by lower bents), and their end as monopolistic units considered above. Odd constructions are zone of high hazard of war.

In this way, monopolistic units and odd constructions are two extreme cases of waning systems whose existing is limited by one level. These cases are linked: monopolistic units multiply without understanding of this process and create odd constructions (may be in their own boundaries), while odd constructions lead to the decreasing of the diversity of their details by the activity of monopolistic units which arise in them and destroy one another without a control of uniting process.

Among concrete systems limited by one level are all onelevel theories which ban to change their laws (axioms). These theories describe real units and acts of diverse strata by axiomatic records which cannot take into account and connect they lower and higher strata (Exact defining of onelevel theories is contained in (Novikava et al 1995)).

The ban of changing of lower levels statutes by acts of multiplying&uniting is the main sign of waning systems. The secondary signs of waning systems are arms (in monopolistic units) and wars (in odd constructions). Waning systems (Fig.5) intend to decrease the level of enemy from within and outside. Because of that they decrease their own levels. But they don't hinder the level increasing process.

Actual sway must change the lower strata units and constructions by the increasing of their levels. That is the sway acts without the arms. There are no arms of lower levels which can destroy the higher strata — actual sway cannot be destroyed by lower strata at all, its changing is its amending, the discovering of wider horizons of its might.

3. CONCLUSION

It is easy to understand that the mechanism of work with hierarchical haze contains known ones which deal with uncertainty (theory of chaos, fuzzy sets and other). Unlike them it allows to realize the level increasing process. It practical significance was examinated above in the investigation of active and waning systems. It is the first time when arms and wars where defined by mathematical records with taking into account their internal and external characteristics in diverse real levels. Hierarchical haze is very convenient in investigation of mind (reason) including the minds of large scale systems. (For instance there are now the images of the ethics defined in the texts of pyramids and The Book of the Dead, Iliada by Homer, pagan magic of Belarus, certain directions of China and Indian knowledge, several onelevel world outlooks). Since the work with haze creates the sway of any unit, the mechanisms of aed theory can have practical significance in phycology which must connect the units constructions with the highest strata of hierarchical time&space.

REFERENCES

Novikava S., et al, (1995). Aed Theory and its Realizations by Hierarchical Knowledge Networks. *Proceedings of Conference Supplementary Ways for improving international stability — SWIIS'95*, Vienna, Austria.

Novikava S., et al (1995). State Design: New Way in Exact Sciences. *Proceedings of Conference Supplementary Ways for improving international stability,* — SWIIS'95, Vienna, Austria.

Novikava S., S.Gancharova, A.Zhybul, et al, (1997). The Statute of Hierarchical Mathematics and Its Cybernetical Maintenance. *Preprints of Eleventh International Conference on Mathematical and Computer Modelling and Scientific Computing.* Washington, DC, USA.

Kile F. (1998). Confronting Cassandra's Dilemma — Sustainability and Peace and Future Leadership, *Proceedings of Conference Supplementary Ways for improving international stability* — SWIIS'98, Romania.

ARTIFICIAL SOCIETIES: A NEW PARADIGM FOR COMPLEX SYSTEMS' MODELLING

Monica Dascalu, Eduard Franti*, Gheorghe Stefan

"Politehnica" University of Bucharest,
Bd. Iuliu Maniu 1-3, cam.413, Bucharest, Romania
**National Institute for Microtechnologies (IMT),*
PO-BOX 38-160, Bucharest, Romania

Abstract: The new paradigm of artificial societies aims to simulate human societies and to study collective (social) phenomena from bottom up. The model of this new paradigm, introduced by Epstein and Axtell, is called "sugarscape". The sugarscape society simulates trade, but is not an industrialised world. The experiment described in this paper considers also production and trade (in a new way): some agents produce a new commodity (vitamins) that enhances agents' visibility for a time, while the other buy this new commodity. Simulations have shown that population dynamics is improved and clearly distinguished social classes appear. *Copyright © 1998 IFAC*

Keywords: social and behavioural sciences, economic systems, simulators, cellular automation, agents, ecology.

1. INTRODUCTION: A NEW PARADIGM

Any major progress of the computer science and computational techniques implies the introduction of a new paradigm. Until now, there are three important paradigms whose sequence describe the evolution of the domain: numeric calculus, artificial intelligence and artificial life. Each paradigm is associated with a computation model and a type of architecture; the adequacy of these elements strongly influences the power of the paradigm. The next paragraphs will describe the paradigms mentioned above and will introduce a new one - artificial societies - that seems to be the natural direction of evolution in computer models.

The initial paradigm - numeric calculus - is well known, as it remained until now the basis of the most used computer systems. It is based on the Turing machine computing model and it uses the classical Von Neumann architecture. The strongest limit of the Von Neumann architecture is the fact that it is basically sequential, and thus the introduction of the parallelism appeared to be the solution to all the deficiencies of this model. However, nobody discovered the way to transform a sequential architecture in a parallel one.

Artificial intelligence, also sustained by the Turing machine model and Von Neumann architectures, as it started from a sequential approach, had few results in improving the computer techniques. The aims of the artificial intelligence are also declared (Winston, 1992) as understanding of the human intelligence mechanisms and principles and give the computers an intelligent-like behaviour, or the study of the computations that make it possible to perceive, reason and act.

Going further, artificial life is devoted to the creation and study of lifelike organisms and systems built by humans. The organisms of this life are not natural, as they are made of nonorganic matter, which essence is information: as Levy (1993) writes, computers are the kilns from which these new organisms emerge. The artificial life paradigm is based on the computation model of Lindenmayer grammars and proposes a model that implies parallelism. but until now there are no adequate architectures. The two important trends of artificial life research use cellular automata and agent modelling.

The new paradigm of artificial societies aims to simulate human societies and to study collective (social) phenomena from bottom up. The model of this new paradigm is introduced by Epstein and Axtell (1996) and is called "sugarscape". This model has a very well suited associated architecture: cellular automata. Basically, cellular automata are lattices of cells evolving synchronously according to identical local rules that depend on local conditions. Appropriate combinations of cellular automata successfully sustain artificial societies models and experiments.

Artificial societies combine cellular automata and agent based modelling, the two major trends of artificial life. Cellular automata often use the notion of signals for homogenous simple agents in simulations of ecological or social systems. The agents of the artificial societies are heterogeneous.

Before describing artificial societies in more detail. let us notice that this paradigm, for the first time. proposes in the same time a model and an architecture and beneficiates of the advantages of their adequacy. This is certainly one of the reasons of the very promising results in artificial societies research, results that recommend them as a powerful scientific tool for social sciences.

The present paper presents a new artificial society experiment. demonstrating once again how this model allows collective significant behaviours to appear through simple individual rules.

2. AN ARTIFICIAL SOCIETY MODEL

Artificial societies are agent-based models of social processes. The artificial society model introduced by Epstein and Axtell (1996) will be briefly presented. as it will be used for the experiments reported in this paper. Epstein and Axtell applied agent-based computer modelling techniques to the study of the human social and economic phenomena such as trade, migration, regional and cultural group formation, combat, interaction with an environment, transmission of culture, propagation of disease, and population dynamics. In this approach fundamental social structures and group behaviours emerge from individuals operating in artificial environments. Both agents (individuals) and the landscape (environment) have simple local evolution laws, thus requiring only bounded demands on each agent's information and computational capacity.

The aim of the research is to grow artificial societies that can be used as laboratories for social sciences and in the same time to discover the fundamental local mechanisms and proprieties that are enough to generate a global (collective) behaviour of a certain complexity. The model involve three basic categories: agents, their environment and the rules of behaviour for both agents and environment.

The former approaches were much simpler, differing essentially from this one by the homogeneity of the agents. Cellular automata based social models, for instance. successfully simulate the behaviour of homogenous collectivities (such as ant colonies) but cannot be directly applied to socio-economical human phenomena.

Human society is one of the most complicate and complex systems that scientist want to understand. explain and predict. Artificial societies' success in modelling socio-economical systems gives a good measure of the power of this paradigm.

2.1 Basic components of an artificial society

The environment is a landscape containing resources (of a single type) that are distributed according to a certain topology. The agents look for resources in the landscape, as they need to eat and metabolise them in order to survive. Thus, the environment is modelled as a lattice of sites bearing resources (practically a cellular automata). it is a medium separate from the agents. on which the agents operate and with which they interact.

The agents are the people of the artificial society. Each agent has internal states and behavioural rules that can change through interaction with other agents or with the environment. Each agent need resources to survive, resources that he has to search in the landscape, eat them and consume them with a metabolic rate. The characteristics of an agent can change in time or remain fixed.

The particular environment chosen for experiments is a "sugarscape": the resource that the agents need to consume for survival is called "sugar" and is distributed in the different sites of the lattice. Each site has a maximum and an actual sugar level (measured in "units"). together with a simple growing-back rule for this resource after one agent have consumed it. The growing-back rule can imply, for instance, that the resources regenerate with one unit per time step or that they regenerate completely in the next time step.

Each agent is characterised by a set of fixed and variable states. For a particular agent, there are some "genetic" characteristics that remained unchanged during its whole life, and there are others - wealth, for instance, and position in the landscape - that can modify. At a certain moment, the agent is located in one site of the landscape. As it need to consume sugar, his simplest behaviour can be described as searching, eating and metabolising sugar. He also can accumulate "wealth" of sugar, but anyway he is continuously walking around and looking for sugar.

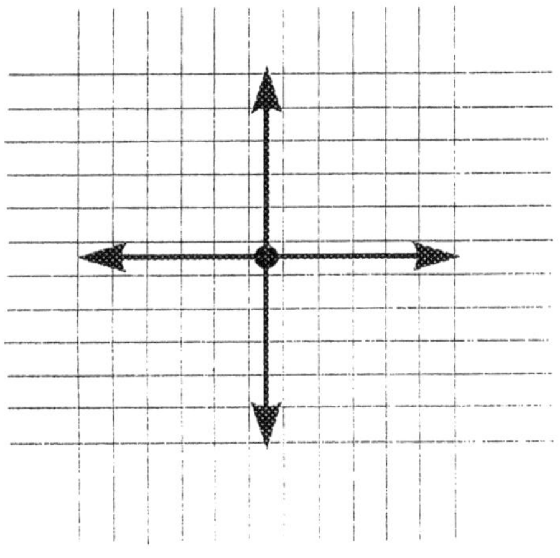

Fig. 1 The agent's vision: the agent sees only in the four principal directions

In this simple description, there are two important characteristics of the agent: a level of vision (how far he can "see" around) and a metabolic rate (the amount of sugar that it burns per time step). Both this values are distributed randomly: the "strongest" agents have high vision level and low metabolic rate, while the ones that have low vision level and high metabolic rate have practically few chances to survive. Figure 1 illustrates the vision of an agent. Note that he can "see" around only in the four principal directions, at a maximum distance equal to its vision.

2.2 Social and economical phenomena modelled with artificial societies

The simplest experiments on the artificial society describe above implies the survival of a population of agents in a certain fixed landscape. In spite of the simplicity of the model, there are some interesting points that can be noticed during the evolution, as for instance the migration phenomenon - the surviving agents move towards the richest zones of the landscape - and the group formation. The agents can live forever if they find the needed resources.

Demography can appear in sugarscape by means of introducing a finite life duration and a kind of sexual reproduction - agents can emerge from the interactions of individuals. In this case, the maximum length of each agents' life can be predefined and, together with the "sex" of the agent, is a fixed characteristic. Each agent has a "fertile" period and it must have a certain "wealth" accumulated in order to have a child. The genetic attributes of the "child" are established after the combination of its parents' and he receives a part of their wealth.

Culture is modelled in the sugarscape by means of a "culture vector" that can modify through local interactions between neighbours - an elementary modelling of the human cultural influences by interactions between people. Although the influences are random, cultural groups emerge, in which all the members have similar cultural attributes.

If instead of one common population there are two types of agents, belonging to two different "races", conflict can be easily modelled by establishing appropriate interaction rules between the members of the two "ethnic" groups.

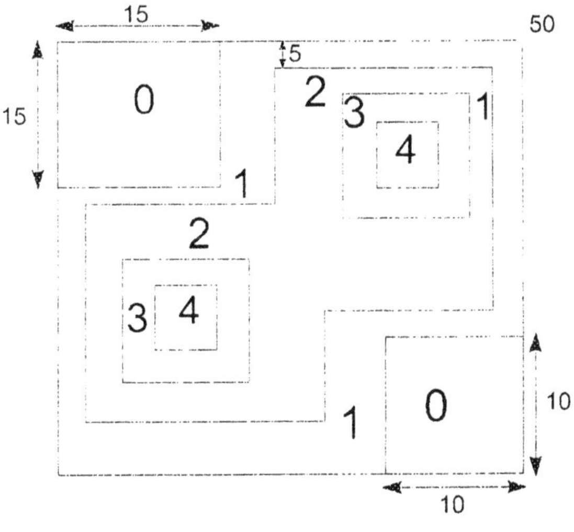

Fig. 2 The resources distribution in the production and trade experiment.

Epstein and Axtell (1996) consider that sexual reproduction, culture and conflict rules are enough to develop a "proto-history" on the sugarscape.

Another crucial social behaviour, trade, can be introduced by assuming two different vital resources (sugar and spice), each distributed in the environment. Agents find, consume and accumulate both of them and can make exchanges with their neighbours to compensate their needs. Finally, disease processes are modelled taking into account disease transmission and immune response.

The conclusion of the studies and simulations of the artificial societies is that important social (collective) phenomena can emerge from spatio-temporal interaction of autonomous agents operating on landscapes under simple local rules. The artificial societies can be used as a scientific tool for social sciences, a domain where experiment seemed to be impossible. The following experiment illustrates the capabilities of artificial societies and the effect of simple individual laws upon the global behaviour.

3. PRODUCTION AND TRADE SIMULATED WITH ARTIFICIAL SOCIETIES

This section describes a new experiment on the artificial societies' sugarscape. The aim of this experiment is to illustrate how a simple new feature can affect the collective behaviour.

Some details about the basic model are necessary. First, the resources of the sites regenerate with one unit each time step, if they are less then the maximum value. The initial resource distribution is illustrated in Figure 2.

The basic agent movement and metabolism rule implies the following:

- look around as far as vision permits in the four principal lattice directions and identify the unoccupied site(s) having the most sugar;
- select the nearest site with the greatest sugar value;
- move to this site;
- collect all the sugar at this new position;
- compute the new value of the agent's wealth, adding the collected sugar and substracting the sugar consumed (equal to the metabolic rate);
- if the wealth is less than zero, the agent die of starvation.

The agents can also die "naturally', when they have attained their lifelong duration. In this case, they are replaced with a new-born agent, having the position and genetic attributes selected randomly

For one time step, the agents act in a random order. Initial positions of agents, their wealth and all genetic treats are randomly distributed within the appropriate intervals.

3.1 Production and trade

Somehow resembling to what happens in a proto-society, the richest agents start to produce a certain commodity, named "vitamin". Production implies a cost: vitamins do not simply appear. a certain amount of sugar is consumed for each unity of vitamin produced.

Producers can sell this product to the other non-producers agents at a cost considerably greater than the production cost: this is trade. It is done on the basis of a request mechanism: if the agent can afford to buy "vitamins" (his wealth is greater than a "poverty threshold"), he looks around to see if there is somebody selling. If there is, he buys as much as he can afford (or the other offers) and consumes all the vitamins immediately.

Vitamin's effect is the following: they improve the agents' skills for a certain period of time. Practically, the agents have a better vision for a period that depends on the quantity of vitamins consumed. Note that this second commodity is not necessary for survival and that it has no conditioning effects on the metabolism.

3.2 Conditions of the experiment

The experiment takes place in a sugarscape of 50x50 sites with the resource distribution in Figure 2. The initial population has 400 agents, having the vision in the interval (1,6) and the metabolic rate in the interval (1,4), initially randomly distributed.

The wealth necessary to become a producer is 60 sugar units; one vitamin is produced at cost 1 and selled at cost 5. It offers a maximum vision for two time steps. The "poverty threshold" is 20 units.

The aim of the experiment is to measure the evolution of number of agents, mean vision and wealth distribution in time, in order to compare the results with the same measurements, on the same population and under the same initial conditions, without vitamin production.

Note: the conditions of the experiment keep as close as possible to Epstein and Axtell's simplest sugarscape, in order to have a reference. Simulations were made with a C++ program; producers belong to a subclass of the general agent class.

Production and trade are performed after the algorithm given above for searching and accumulating resources, practically after the agent is placed in a new position.

3.3 Measurements and results' interpretation.

The initial phase of evolution implies group formation, while agents move to the richest zones in sugarscape. Many agents die of starvation, and the final population has a stable value, named carrying capacity.

Figure 3 and 4 compare the population dynamics, the mean vision evolution in simple conditions and with production and trade. Figure 5 gives the typical wealth histogram in the two cases, after 500 time steps.

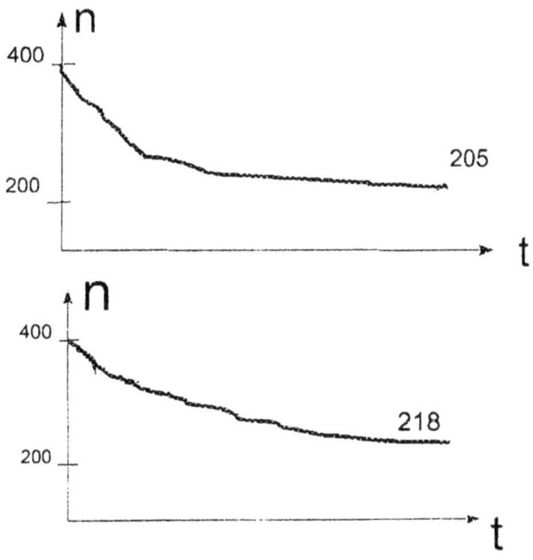

Fig. 3 Population dynamics: the evolution of the number of agents (n) in time, for simple sugarscape (up) and production-trade experiment (down).

Running the new artificial society scenario, one will notice the following aspects:

- the vitamins modify the dynamics of the agents in the sugarscape and also the population dynamics;
- the effects of the production and trade on the population dynamics in the first period of the experiment, when agents are still looking for the best
- place to live, is to help agents to survive;
- in the stable period of the evolution, the carrying capacity is greater, due to the increased mean vision;
- in the stable period of the evolution, the wealth distribution is seriously affected, as agents do not really need to be good "food searchers", but they still buy vitamins;

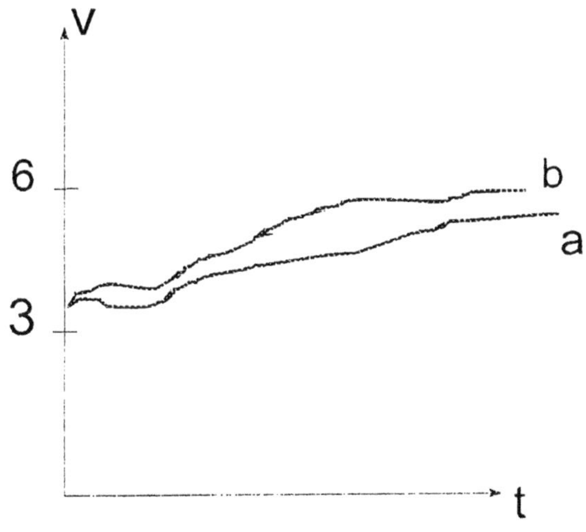

Fig. 4 Mean vision (v) versus time in simple sugarscape (a) and production experiment (b).

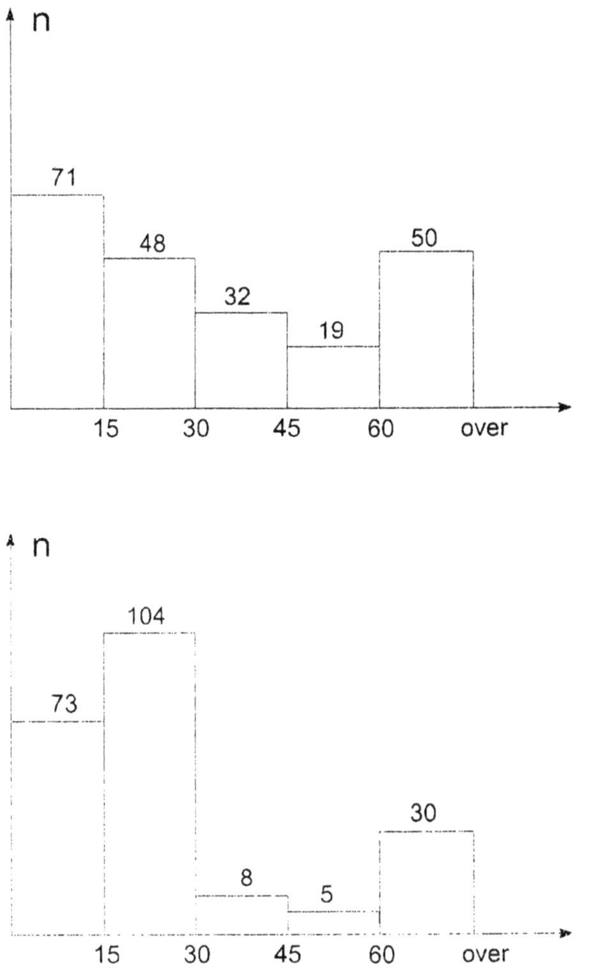

Fig. 5 Typical wealth histograms for simple sugarscape (up) and production experiment (down). Note the clear "class" formation.

- producers become richer and richer as time goes by;
- the agents in the medium wealth category are mentained at the "poverty threshold" and do have to worry for tomorrow, as they do not have some accumulated reserves;
- for the poor, nothing change.

The crucial new phenomenon that appears is social class formation (after the agents' wealth). Producer's number remain unchanged, as the medium class agents cannot reach the required wealth to become producers.

3.4 Some comments

All these aspects are quite similar to what happen in real societies, if the things are putted crudely this way: there are some products which make people's life longer and easier and these products have a good effect on the global dynamics of population. Unfortunately, people consume these "good and necessary" products even if they do not really need them. As a result, producers become richer and richer while people with medium wealth spend their resources on futile things. The poor are just poor all the time.

A more realistic approach (and a further experiments) needs some special new trade rules. in order to limit the quantity of vitamins consumed by non-producers agents. The agents of the medium class being "greedy", they will always remain close to the poverty threshold.

4. CONCLUSIONS

Artificial sciences, a new paradigm for complex systems modelling, deals with socio-economical phenomena combining two "instruments" of artificial life: cellular automata and agent modelling.

Simple landscape plus simple agents can produce a complex collective behaviour, well suited to model real economical and socio-cultural structures.

In the production-trade experiment, the social class structure that emerge through simple rules is remarkable, together with the increase of the carrying capacity and mean vision (that reflect the "good" effect of the new feature, production). As this experiment was intended to illustrate the power of simple rules in the artificial societies model, the results are clearly positive.

REFERENCES

Epstein, J. M. and R. Axtell (1996). *Growing Artificial Societies - Social Science from the Bottom Up*. Brooking Institution Press. Washington D.C. The MIT Press, Cambridge. Massachusetts & London, England.

Steven Levy (1993). *Artificial Life - the Quest for a New Creation*. Penguin Books. London. England.

Patrick H.Winston: *Artificial Intelligence*. Third edition. Addison-Wesley Publishing Company, 1992

STABILITY VERSUS UNCERTAINTY

Vlad Ionescu

Faculty of Automatic Control and Computers
University "Politehnica" Bucharest

Abstract: In this paper an attempt has been made to describe, in state-space or operator based description, the leading ideas that govern the system evolutions. The importance of the dichotomy property it is also emphasized. Some considerations on robustness are given as well. *Copyright © 1998 IFAC*

Keywords: stability, dichotomy, generalised distance.

1. INTRODUCTION

It is a matter of consensus that the notion of stability, originated rigorously by the Systems Theory has presently various connotations. However it hides the same meaning: the tendency of a given system to approach an equilibrium state. From a philosofical perspective, at least three questions could be naturally posed:

a) is stability an intrinsic property?
b) can the system be stabilized? If the answer is positive then provide an appropriate explicit procedure for achieving stability.
c) once the system made stable try to evaluate a stability margin with respect to the parameter uncertainties.

In what follows we shall restrict our attenntion to the linear discrete case and we shall try to discuss the above mentioned aspects.

2. STABILITY: AN INTRINSIC PROPERTY?

Let us consider the evolution generated by the *direct* time:

$$x = (A\sigma^{-1})x := A(\sigma^{-1}x) \qquad (1)$$

or reverse time recursions

$$x = (A\sigma)x = A(\sigma x) \qquad (2)$$

Here A is an l^2 - bounded multiplication operator, i.e. $A = (A_k)_{k\in Z}, A_k \in R^{n\times n}, \|A_k\| \le c_0 \forall k$ and σ is the unit forward bilateral shift, that is, $(\sigma x)_k = x_{k+1}$.

Hence (1) and (2) yield explicitly

$$x_k = A_{k-1} \dots A_i x_i, k > i \forall k, i$$

and

$$x_k = A_k \dots A_{i-1} x_i, k < i \forall k, i$$

respectively. One can easily see that both $A\sigma^{-1}$ and $A\sigma$ are l^2 - bounded linear operators. We shall say that (1) ((2) defines an *exponentially stable* (ES) *(anticausal exponentially stable* (AES)) evolution, see Halanay and Ionescu (1994), if

$$\rho(A\sigma^{-1}) < 1 \quad (\rho(A\sigma) < 1) \qquad (3)$$

Here ρ stands for the spectral radius. If we relax the above inequality to $\rho(A\sigma^{-1}) \le 1 \quad (\rho(A\sigma) \le 1)$ we shall be confronted with the *semistable* (anti)causal evolution. Clearly if $\rho(A\sigma^{-1}) < 1 \quad (\rho(A\sigma) < 1)$ then $x_k \to 0$ as $k \to \infty$ ($x_k \to 0$ as $k \to -\infty$). In the semistable case the boundedness of the evolutions is guaranted only, i.e. there exists c_0 such that $\|x_k\| \le c_0 \|\xi\| \quad k > i \quad (k < i)$ for any initial state ξ at $i \in Z$. Taking into account the above definitions we may conclude that ES (AES) *is not generically fulfilled*. The property depends drastically on the internal parameters, hence it depends on the influence of the uncertainties. A comment which naturally arises is the following. Let A be given and let Δ be any l^2 - bounded multiplication operator. We shall say that Δ acts as multiplicative uncertainty if A is modified in $A_\Delta = A(I+\Delta)$. Clearly if A is ES (AES) then the spectrum Λ of $A_\Delta\sigma^{-1}$ $(A_\Delta\sigma)$ may be spread in the whole complex plane. However we can admit that for almost all Δ we have

$$\Lambda \cap T = \Phi \qquad (4)$$
$$(T = \{\lambda : |\lambda| = 1\}).$$

Can we describe such situation? In this way we are led to the notion of *exponential dichotomy* (ED), see Ben-Artzi and Gohberg (1989) and Halanay and Ionescu (1994). We shall say that A defines an ED evolution if exists a Lyapunov similarity such that

$$A \approx diag(A_1, A_2) \qquad (5)$$

where A_1 defines an ES evolution, A_2 has a bounded inverse and in addition A_2^{-1} defines an AES evolution. In fact (5) show that two evolutions occur at the simultaneously: one in direct time and one in reverse time, i.e. $x_1 = A_1\sigma^{-1}x_1$ and $x_2 = A_2\sigma x_2$. Now we can remark that the bounded invertibility of A_2 shows that the second system is time-reversible - a condition that often could be considered to be too strong. In order to relax such assumption the evolutins (1) and (2) will be both inbedded in a more complex one, namely the evolution of *descriptor* form, i.e.

$$Mx = N\sigma^{-1}x \qquad (6)$$

Systems of form (6) are usually encountered in economical or social processes. The notion of ED is introduced as follows: $M - N\sigma^{-1}$ is said ED if there exists two Lyapunov transformations U and Z such that

$$U(M - N\sigma^{-1})Z = diag(I - S\sigma^{-1}, T - I\sigma^{-1}) \qquad (7)$$

where S and T define ES and AES evolution, respectively. Here $S = (S_k)_{k\in Z}, S_k \in R^{r\times r}$ and r is termed as the *rank* of the dichotomy.

Based on the above considerations one concludes that neither stability nor antistability are intrinsic generic properties of a system. The most natural property that a system possesses generically is that of dichotomy.

3. ABOUT STABILIZABILITY

In Systems Theory the notion of stabilizability has a very exact meaning. It refers to the fact that a system can be made ES (or AES) via an appropriate state-space feedback law. in this case the evolutions (1) and (2) must be directly influenced by o control input. That is, they become, in descriptors form

$$Ex = A\sigma^{-1}x + Bu \qquad (8)$$

and

$$Ex = A\sigma x + Bu \qquad (9)$$

respectively. Thus if the state space feedback laws $u = F\sigma^{-1}x - Gx$ and $u = F\sigma x - Gx$ are performed on (8) and (9), respectively, one gets:

$$(E+BG)x = (A+BF)\sigma^{-1}x \text{ and}$$
$$(E+BG)x = (A+BF)\sigma x, \text{ respectively.}$$

Clearly it is quite transparent that (8) ((9) is stabilizable (antistabilizable) if:

$(E+BG)-(A+BF)\sigma^{-1}$ $((E+BG)-(A+BF)\sigma)$ admits a dichotomy of *full (null)* rank. However x is not ussualy accesible for measurement. Hence we need to get information about the internal evolution through a measured output. Thus, for instance (8) must be considered in the augmented form

$$\begin{aligned} Ex &= A\sigma^{-1}x + Bu \\ y &= C\sigma^{-1}x + C_1x + Du \end{aligned} \qquad (10)$$

If $E - A\sigma^{-1}$ admits a dichotomy, then (10) defines an l^2 - bounded input-output operator

$$T = (C\sigma^{-1} + C_1)(E - A\sigma^{-1})^{-1}B + D \qquad (11)$$

Such an operator is said *contractive* if $\|T\| < 1$.Of major importance is a *Small Gain Theorem* which asserts the following. Let

$$\begin{aligned} E_ix_i &= A_i\sigma^{-1}x_i + B_iu_i \\ y_i &= C_i\sigma^{-1}x_i + C_{1i}x_i + D_iu_i \end{aligned} \qquad (12)$$

$i=1,2$ be two dichotomic systems with the dichotomies of rank r_i, $i=1,2$ be the appropriate input-output maps defined via (11). Assume that both systems are contractive and feedback compatible. Then the closed loop configuration is dichotomic with the rank of dichotomy r_1+r_2. Hence if the dichotomies are of full rank then the closed loop configuration is ES.

Such result reveals the influence of the uncertaities Δ_1, Δ_2 , i.e. as long as $T_i(I+\Delta_i)$ $i = 1,2$ are contractive, the closed loop configuration preserves the dichotomies and cummulates the ranks. Particularly the ES is preserved.

4. STABILITY MARGIN. ROBUSTNESS

Let (10) be given. Assume that it is ES. For a given tolerance $\gamma > 0$ find an AES S such that

$$\|T + S\| < \gamma \qquad (13)$$

Clearly the least achievable tolerance γ_{min} is an evaluation of the robustness of (10). It can be proved that it equals exactly the Hankel norm of T and it is seen as a *stability margin* .

5. CONCLUSIONS

In this paper we tried to give an overlook on the basic notions related to the concept of stability in large sense. Evolutions of descriptor type have been considered under the dichotomy hypothesis. Some considerations on the stability margin are also given.

REFERENCES

Ben-Artzi, A. and I. Gohberg (1989), " Inertia theorems for nonstationary discrete systems and dichotomy", *Linear Algebra and Its Applications* **120**, pp. 95-130

Halanay, A. and V. Ionescu (1994), *Time-varying discrete linear systems*, Birkhauser

GEOPOLITICAL RELATIONS IN POST USSR EUROPE AS A SUBJECT OF MATHEMATICAL MODELLING AND CONTROL

S.Levkov*, A.Makarenko**

* *New Jersey Institute of Technology, ECE, University Heights, Newark, NJ 07102, USA. E-mail: levkov@megahertz.njit.edu*

** *National Technical University of Ukraine (Kiev Politechnic Institute), MMSA Dept.,19-20, 37 Pobedy Ave, Kiev, Ukraine*

Abstract: A mathematical model describing dynamics of geopolitical interrelations is considered and illustrated on an exmaple of the model of evolution of the bipolar GPS in Europe. The integrated characteristic of a country's power is taken as the state variable. Description of the system dynamics is based on the principles of associated memory models. The matrix of inter-state connections is constructed upon historical patterns and geographic factor. The comparative quantitative results are obtained on the computer model. *Copyright © 1998 IFAC*

Keywords: socioeconomic system, geopolitical relations, modeling, associative memory

1. INTRODUCTION

The problems of international stability are of great theoretical and practical interest. Especially important these problems became recently after the disintegration of the USSR and resulting changes of bipolar geopolitical structure. Many famous scientists considered these problems from different points of view (Ratzel, Guetzkow, Galtung, Morgentau, Aron, Frank, Vallerstign, Hantington, Rotfeld and others). Lately the methods of mathematical modeling and adaptive control were developed for such problems (see the works of P.Kopacek).

The analysis of international stability requires a global approach (see f.i. its connection with the problems of sustainable development (Klestova, 1997). Despite the existence of numerous global models (f.i. the famous models of J.Forrester and J.Meadous), the adequate modelling approaches are yet to be developed. The approach adopted here may help to consider new type of global models that take in the account a notion of global culture of society, and where the international relation and stability would be only one of model's sub-blocks.

International stability is strongly connected with development of geopolitical structure (blocks, treaties, unions, their relations) in current and future World. Considering the geopolitical structure (GPS) as a dynamic system we can obviously conclude that at the time being this system is at the unstable point and begins its transition to may be another stationary state. What will be this future state? Will it be the uniform community of western like democracies, will it return to previous structure similar to NATO - Warsaw Treaty withstanding or will it be the unknown multi-polar structure?

There are several reasons that make the answer not so easy to find using traditional methods of intuitive expert estimations. The major one is that the current system of GPS, being evidently the typical example of large scale system, is becoming more and more complicated involving too many countries whose strategic interests now are not well defined and stable. Second, the economic situation is deteriorating in most of the X-USSR countries as well as in some other developing countries. Third, the economic crises in X-USSR states is aggravated by psychological crises and deceived expectations of people whose mentality was grown under communist ideology. Last but not least is the epidemic of ethnic conflicts which is spreading across the whole World.

All this leads to a natural desire to have at hand some instrumentation which would help in estimating the probable changes in GPS, i.e., the mathematical model. The mathematical methods traditionally are not widely used in such fields as international relations and political sciences, but recent expansion of mathematical modeling, synergetic and cybernetic in economics, biology and other nontraditional directions allows to expect a positive result in our case.

An approach is proposed here to build a model for description of GPS. The method is based on the idea of A.Makarenko (Makarenko, 1993), (Danilenko et al., 1992) to use the synergetic concept to describe in one model the social, economic, ethnic, behavioral and historical processes in society. In this first attempt the model of the evolution of bipolar GPS in Europe is presented. The integrated characteristic of a country's power is taken as the state variable. Description of the system dynamics is based on the principles of associated memory models. The matrix of inter state connections is constructed upon historical patterns and geographic factor. The quantitative results are received on the computer model in the case of two, three and 2N+1 levels of every state.

2. SOME CONCEPTS OF GPS IN CURRENT WORLD

Before developing the concept for the mathematical model of GPS it is useful to consider briefly the major existing ideas and hypothesis in the field of geopolitics. We will be interested here mainly in the historical period beginning from 40-50 of XX century.

According to the common point of view after the World War II the world was decided into East and West major global blocks with the USSR and USA as dominating poles (Freadman, 1992), (Jackson, 1992). This GPS (called bipolar) with all its inherent defects proved to be stable enough and provided collective security due to the balance of force and strategic weapons.

The events of last decade broke the stability of bipolar GPS. The process of its significant transformation was triggered by several factors: disintegration of USSR, the effect of mutual peaceful initiatives, the changes in the balance of force inside the Western block and emerging of new political agents among the developing countries. In what direction these changes will lead the international community? What will be the future most probable GPS? These questions are not only of pure academic interest, they have also the great importance in evaluating the consequences of practical steps which influence the creation of new GPS.

Presently several concepts of future world structure exist, some of which are incompatible. Thus in (Klein, 1990) several possible alternatives are described including the world community without state structure and on the other hand over national structures. Concerning the traditional structures the multipolar constructions are often considered (Freadman, 1992), (Klein, 1990) where the attraction centers can be different from traditional ones. Thus in (Lukin, 1983) the idea is considered about the three level differentiation of world countries: leading, developed and passive ones.

According to R.Klein (see (Lukin, 1983)) the world will be divided into several regions more or less independent in their development. The multipolar concept is also considered, where the problem of main direction of concentration of contradictions (East-West or North-South) is most controversial. Instead of bipolar GPS the idea of two major power regions is sometimes considered. The three polar structure for western world was suggested in (Freadman, 1992),(Jackson, 1992). In (Jackson, 1992), different constructions like global village model, democratic model, collective security model, pax-American model are discussed also.

One-polar structure is considered in (Round table, 1991) where the center is understood as pole consisting from three most developed countries. Such construction can be derived from existing 7+1 configuration. Quite different to this, the integrated structure based on horizontal connections is considered in (Gadgiev, 1993).

One more possibility cannot be disregarded, the one without stable blocks and stable structure at all. In such a case the world can become less stable and the concept of local balances of power and interests can lead not to the world stability and security (Gadgiev, 1993).

The straightforward resume from this brief analysis can be the following. The potential multivariance of future GPS makes strongly desirable modeling the GPS evolution, the possibility of its control being considered as the major incentive for such modeling.

3. FORMAL DESCRIPTION OF GEOPOLITICAL RELATIONS

Most of the notions in the field of geopolitical relations have usually only verbal definitions. The problem is to find the models which would allow to give the qualitative meaning to them.

First we have to define the set of elements which will be the objects of relations. Quite obviously these could be the independent states. Sometimes, when the future processes of potential disintegration are to be taken into the account the separate regions can be considered also as constructs for building the model.

Second, the problem of formal description of characteristics of the elements arises. The basic one of them is the notion of power of a state. Significant studies were performed in (Klein, 1990), which showed that power incorporates such concepts as population, area, geography, economy, political factors and military aspects.

R.Klein suggested the integrated characteristic for the state power

$$P = (C + E + M)(S + W + F) \tag{1}$$

where P- cumulative power of a state, C- critical mass (the sum of coefficients of population and area), E - economical power, M- military power, S- strategic goals of a state, W - desire of population to follow the existing strategic goal, F- convincing force of political leader, his ability to lead the own population and aliases. All the parts of the formula 1 are estimated usually by experts.

Following this work we can also introduce an integrated characteristic of a single state which we note as s. Since we are interested in dynamic modeling we will consider it as a function of time $s(t)$ in continuous case, or $s(kt), k = 0,1,2...$ in discrete case. Usually the estimates are obtained through fixed time intervals t : $s(0), s(t), ..., s(kt)$ where most frequently t represents one year. It is clear that such description of the state is essentially rough. More exact solution consist of introducing the vector of state parameters such as population, economic power, military power, strategic weapons etc.

Thus we consider N states with the state variables $s_i(t)$, $i = 1...N$. The next step is to formalize relations between them. With the states i and j we connect the variable J_{ij} - integral force of connection of state i with state j (in general case it is not supposed that $J_{ij} = J_{ji}$). In the case of scalar $\{s_i\}$ this connections are integral characteristics and include economic, military, ideological and other connections. They can be positive and negative as well (for instance if the bipolar GPS is the case the connection between the states from different blocks can be obviously negative what means the repulsion between such states). In the case of vector state variables $\{s_i\}$ the vector of connections between states i and j should be considered. The set of connections J_{ij} between the state of region considered can be defined with the help of expert estimates or be calculated according specific rules which will be considered below.

Generally speaking connections depend also on the time, on vector of parameters μ, and on space coordinates x (geographic position of a state):

$$J_{ij} = J_{ij}(t, \mu, x) \tag{2}$$

The next step consist of deriving the formal laws governing the dynamics of the system. In general, these laws must depend on the state variables of the states, connections between them, parameters. This dependence can include the certain period of history and can be described in the following general form:

$$\frac{ds_i}{dt} = f_i(\{s_i(t)\}\ ,\{J_{ij}(t)\}, b) \tag{3}$$

or

$$\begin{aligned} s_i(t+1) = \varphi_i(\{s_i(t)\}, \{s_i(t-1)\}, ..., \\ \{J_{ij}(t)\}, \{J_{ij}(t-1)\}, ..., b) \end{aligned} \tag{4}$$

where f_i - may include integral dependence, b - some additional parameters.

In general case we cannot expect that connections between states can be kept unchanged in the time. Thus we may introduce the formal laws for the connections dynamics in the form 5 or 6:

$$\frac{dJ_{ij}}{dt} = g_{ij}(\{s_i(t)\}\ ,\{J_{ij}(t)\}, \mu) \tag{5}$$

$$\begin{aligned} J_{ij}(t+1) = \psi_{ij}(\{s_i(t)\ \}, \{s_i(t-1)\}, .., \\ \{J_{ij}(t)\}, \{J_{ij}(t-1)\}, ..., \mu) \end{aligned} \tag{6}$$

In the most general form the problem of modeling and predicting dynamics of GPS includes full description of dynamic laws both for $s_i(t)$ and $J_{ij}(t)$. Different time scales make possible to consider in some cases the problem of evolution of GPS under the fixed connections ($J_{ij}(t) = const$). But in any case the description of dynamic laws is a key factor in building an adequate model. Some ideas about constructing such laws are discussed below.

4. ASSOCIATIVE MEMORY APPROACH TO BUILDING THE DYNAMIC MODELS OF GPS.

Our approach is similar to the class of models proposed in (Makarenko, 1993),(Danilenko et al., 1992) and is based on the following considerations, which have the source in the analysis of current state and history of international relations.

It is almost obvious that connections between states are not accidental but flow out from the historical process of international development. It is well known that there existed in history well defined stable associations. The membership of a state in such associations is determined by some inner properties of the state and vice versa this membership influence itself such properties. Thus every new GPS is supposed to inherit the historical traces.

The change of components of the state variables vector can be caused by two group of reason: inner and outer. For instance the state economics is changing mainly under the influence of internal factors, though external financial and technological factors also have the importance. The same is with military power, strategic goals, etc. On the other side, population is an example of the variable influenced mostly by internal factors. In this work we restrict ourselves in considering only exogenous component of influence. This may be interpreted as taking the projection of state variable on external components. Thus zero value of s_i does not mean zero state power but rather its neutral character, i.e. non influence on some external projection of other states. So in this context the state variable s_i means the power related to some union or block.

Let us consider now the potential type of laws which could describe dynamics of $\{s_i(t)\}$. The external component of influence depends on the state of other countries and relations with them, i.e. on the great number of different forces. Thus we can suggest that this influence can be described by some average value, like field in physics. We can take for the value of this field of influence on i-th state the weighted sum of state variables:

$$W_i = \sum_{j \in U_i} J_{ij} s_j \qquad (7)$$

where U_i- the surroundings of i-th state. In the case of fixed (or external to the model) interstate connections, the dynamics of $\{s_i(t)\}$ in laws 3, 4 will be defined as first approximation by expressions of type 7. The description of GPS in the terms of $\{s_i(t)\}$ and $\{J_{ij}(t)\}$ together with above said reasons allows to find out among the possible dynamic laws the most suitable ones. This is based on the use of results of current research in the field of system analysis in cybernetics, physics, biology. We mean here the models of self organization. The above analysis allows to make the conclusion that our model is close to the so called models of associative memory which are extensively studied in the problems of pattern analysis (Kohonen, 1980).

Such models consist of great number of elements designed for recognition of some patterns by the mean of association, i.e. evolution of elements to the states related to some patterns which are the attractors in the phase state space. The memorizing of patterns is done by the mean of modification of connections between elements. The simplest model of such kind is the Hopfeild model (Hopfield, 1982), where the every element of the net is described by two possible states +1 or −1, field of influence is taken in the form of 7 and evolution of the state variables is described by the following law (N is the number of elements):

$$s_i(t+1) = sign(\sum_{j=1}^{N} J_{ij}(t) s_j(t)), \qquad (8)$$

Such models possess the ability of learning which allows them to remember the patterns. Thus, if there exists the set of the states of the net $V_1, .., V_k$, $V = \{s_1^l, ..., s_N^l\}$, $l = 1, ..., k$ ($k << N$) and if the inter element connections $\{J_{ij}(t)\}$ are taken in the form

$$J_{ij} = \sum_{m=1}^{k} s_i^m s_j^m, \quad \{s_j^m\} = V_m, \quad m = 1, ...k \quad (9)$$

the system acquire the ability of recognizing the patterns $V_1, .., V_k$. Returning to the model of GPS we suggest to use the principles of associative memory models for the description of the system dynamics. In this case the patterns correspond to the stable global GPS existed in the historical process, description of characteristics of states and theirs connection is taken as in the above paragraph, the dynamics laws are taken the same as in associative memory models. The inter state connections can be evaluated according the formula 9 or based on the method of expert estimates.

5. GPS IN POST USSR EUROPE AS AN EXAMPLE OF EVOLUTION OF BIPOLAR GPS.

The above described concept is implemented here to the modeling of GPS evolution in the post-USSR Europe which is currently in the transition period of its history. As the first approximation we consider here the most simple models.

Model 1. Let us consider the problem of short time prediction. We take the connection $\{J_{ij}(t)\}$ to be fixed and state variables to be scalar. Taking into the account the recent bipolar structure of the Europe we can use the simplest Hopfeild model where $s_i = +1$ means that state is a NATO member and $s_i = -1$ is a Warsaw Pact member. The neutral states are not taken in to the account. As historical patterns, two bipolar GPS are considered: first is the NATO vs. Warsaw Pact, second: Entente vs. Germany and Austria before I World War. For the purpose of modeling the computer program was developed. On the first step of computer experiment the matrix of connections $\{J_{ij}\}$ is evaluated based on the given patterns. This matrix is used then for the calculation of vector $\{s_i(t)\}$ evolution. Experiments consisted in calculating the system evolution to final stable state starting from different initial states. The obtained results showed that such model is far too simple for further research and can serve only as proof of correct work of the algorithm. The model does not have stable states different from given as patterns which are the final points of evolution from any initial state.

Model 2. Is the modification of model 1 where the neutral states are introduced (which obviously take the value 0). The law of dynamic must be also modified. Instead of 8 we use here

$$s_i(t+1) = F(\sum_{j=1}^{N} J_{ij} s_j(t)), \qquad (10)$$

$$F(x) = \left\{ \begin{array}{l} +1, \ x > \Theta \\ 0, \ |x| \leq \Theta \\ -1, \ x < \Theta \end{array} \right\} \qquad (11)$$

where Θ - is the transition barrier or degree of inertia of the state for which several different expressions where studied.

The same computer experiments as for model 1 where accomplished. The results show that the model 2 is substantially more rich in behavior. It can have the stable states different from patterns which depend on initial state and transition barrier Θ. The different examples of evolution starting from the different post-USSR GPS where examined which showed that the pattern 1 and GPS close to it are more attracting then others. To allow the states which were initially in the Soviet block to drift to the West or remain neutral the relatively great values of transition barriers Θ have to be admitted which were close to the values prohibiting any evolution in the system.

Model 3. This model is the modification of Model 2 with some scales for the parameters of states (typically in computations we consider 11 states of every element). There were considered some possible configurations for the future Europe. The experiments on the model with different initial conditions and parameters can show, for example, that Ukraine has a undefinite status between NATO and Russia. This is caused by the balance of forces of surrounding states. This balance caused the series of steps in Ukrainian foreign policy in both (NATO and Russia) directions which neutralized each other. It should be mentioned that if the inertia parameter in model is decreased than there is possible state configuration in Europe reminding the "MittelEurope" configuration with two blocks - first around Germany and second of Entente type. From the model experiments it is also easy to understand such trends as recent treaty between Russia and China as simple reaction of Russia on decreasing of their connections with East European countries, as attraction of Austria to NATO and Belorussia to Russia. Calculations also show as a possibility the strong consolidation of enlarged NATO in one block with neutral or almost neutral status for former USSR countries. This is because the neutral state is hardly changeable. But if the initial state of former USSR countries becomes less neutral, then it may lead to the stronger polarization in future Europe but with enlarged NATO. Some of above mentioned results were discussed at the meetings of Atlantic Council of Ukraine.

It is necessary to mention that the results of modeling being preliminary, show nevertheless the potential possibilities of method considered. Thus, for example, the model allow to calculate the vector of forces acting on a specific country, the creation of new blocks and poles in geopolitics, the relative power of blocks and many others. In the future, the model, enhanced by enlarging its state space, introducing the vector characteristics for single country and by using the expert estimates for determination of interstate connections, can allow to expect the reliable results of predicting GPS evolution in Europe. Another ways of model enhancement are consideration of the elements with inner structure, nonsymmetric models with nonsymetric connections, models with memory effects,and nonstationary bounds, and so on. Proposed approach also allowed to create some new theoretical concepts for the conflict resolution problems ((Makarenko, 1997a), (Makarenko, 1997b)).

The conclusions that can be derived from the experiments with this model of first approximation are briefly: 1) the bipolar GPS that existed in

Europe before the split of USSR is a very stable stationary state attractor (SS^*); 2) Russia is an obvious center of attraction of East European block; 3) even after significant perturbations of this GPS it returns to (SS^*); 4) to prevent the system from returning to (SS^*) a very significant barriers must be introduced into the equations of system dynamics for those countries who are drifting toward neutral state.

The interpretation and analysis of experiments with such models can be the subject of discussions with experts, but preliminary judgments can be uttered already here: to prevent the return of (SS^*) in European GPS the barriers Θ have to be increased.

6. CONCLUSIONS.

The problem of modeling the evolution of GPS is considered in this work. The formal description of GPS is done and the general principle of modeling method is proposed. The analysis of existing and historical GPS allowed to use as the basic class of models the associated memory models. The proposed concept is illustrated on the example of modeling the evolution of GPS in the post-USSR Europe.

It should be mentioned that proposed models are closely connected with modern approaches in self organization theory and synergetic. Such models can serve not only for the purpose of predicting the evolution of GPS but for the purpose of control where two problems may be considered: 1) having the desired GPS, find the system of interstate connection which would make such GPS stable; 2) having the current GPS, find out what change of connection would allow to obtain new stable GPS.

7. ACKNOWLEDGMENT

The authors express their gratitude to Dr. B. Friedland for the support of the work and to A.Dobronravov, D.Nikshich, M.Plostak for the help in computer modeling.

REFERENCES

Bystrov, H. (1981). Methods of estimations of state power (review of foreign military specialists). *Foreign military review* **9**, 12–15.

Danilenko V.A., Korolevich V.A., Makarenko A.S. and Cristenyuk V. (1992). *Self organization in strongly nonequilibria media collapses and structures.* Kiev.

Freadman, L. (1992). Order and disorder in the new world. *Foreign Affairs* **71**, 20–37.

Gadgiev, K. (1993). From bipolarity to new configurations of geopolitical structures. *International economic and international relations* **7**, 62–67.

Hopfield, I.J. (1982). Neural networks and physical systems with emergent collective computation abilities. *Proc. Nat. Acad. Sci. USA* **79**, 2554–2559.

Jackson, R.J. (1992). *Europe in transition. The management of security after the cold war.* Praeger. New York.

Klein, B. (1990). Out of the limits of strategies. *Working class and current world* **6**, 54–66.

Klestova Z., Makarenko A. (1997). New Class of Global Models of Associative Memory Type as a Tool for Considering Globas Environmental Change. *Proceed. NATO Advanced Research Workshop : Environmental Change, Adaptation and Security* Budapesht, Hungary, October 1997.

Kohonen, T. (1980). *Associative memory.*

Lukin, V. (1983). *The centers of power - concepts and reality.* International relations. Moscow.

Makarenko, A. (1993). *A Model of Global Socio-economic processes. Basic Concepts.* Kiev Polytechnic Institute, Preprint. Kiev.

Makarenko A. (1997). Toward the possible origin of life-cycles phenomena in global socio-economical and organisational systems. *Proceedings Int.Conf. Life-Cicle Phenomena, Budatesht, July 1997,* 473–478.

Makarenko A. (1997). Global social conflicts and their models. *In: Conflictological Expertise: Theory and Methodics* 83–90.

Round.table (1991). Microsystem approach today. *International economics and international relations* **11**, 57–69.

DEMOGRAPHICAL UNITS IN HIERARCHICAL COORDINATES: CONSTRUCTION, ACTIVITY AND GOALS

W.Nowik*, V.Barkalin, W.Zianevich***, S.Gancharova****,P.Buka****, K.Miatliuk****, A.Astrowskaya***, A.Bahutski****

**) Belarusan National Academy of Sciences*
***) Institute of Mathematics&Cybernetics*
****) Belarusan University of Culture*
*****) Hierarchical Multilevel Systems Laboratory,*
Minsk, Belarus, 220064, p.o.box 48, e-mail: imc@org.by

Abstract: The paper deals with mathematical definition of demographical units (men, families) in hierarchical coordinates which allow one to connect the lower strata of their internal constructions (physical, chemical and biological layers) with their own stratum as well as with their higher strata (engineering and knowledge systems) which include demographical units and authorize their goals. The definition allows to work out the data bases for demographical strata which are convenient not only in sway (design&control&learning) tasks for demography but for engineering and knowledge levels also. *Copyright © 1998 IFAC*

Keywords: demographical systems, data base, sway networks, hierarchical mathematics

1. INTRODUCTION

The present state of demographical strata is far away from desired one throughout the world (Caring for the Future.,1996). To design the desired state and control the process of its achievement these strata must be studied more detailed, and knowledge about them must be settled in accordance with the requirements of sway (design&control&learning) systems. Besides, the requirement of concrete cybernetic means (Internet among them) must be met since the using of them increases the successfulness of sway activity including the activity of sway networks (Groumpos et al.,1997).

In agreement with hierarchical mathematics (Novikava et. al., 1997) the knowledge construction within the sway process must connect the following details for any system:

- description of its own construction;
- its image as the whole unit in its environment;
- images of its holding systems.

Named details are the hierarchical coordinates of a system of any strata (levels). Without the loss of generality, the coordinates of demographical units in hierarchical space may be made more concrete if the concrete construction of this space suggested in (Novikava et. al., 1995) are taken into account. This construction contains the following strata: natural (physical, chemical, biological), demographical (men, human populations, families), engineering (industry, service, conveyance, trade, monetary mechanisms), knowledge (languages, art, design, scientific and learning institutes). In this approach the demographical units contain natural strata as their own constructions and can exchange with other

$^{+}\alpha^{\lambda}$:

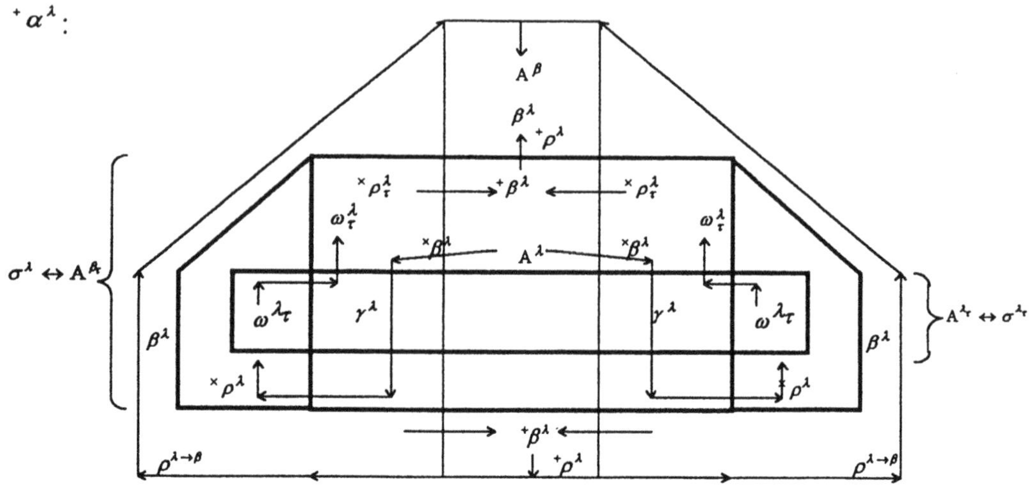

Fig. 1 Graphical symbol-image of aed (the scheme of aed processor).

units of their own level by the details of these constrictions. This exchange is the base of definition of them as active units which create jointly their sway (holding systems) - the engineering strata - and reform knowledge one. Holding systems have the ranges which contain diverse demographical units whose goals and psychological characteristics are cohered with their movements in these ranges.

In accordance with that the paper includes:

- abridged description of laws and symbolism of hierarchical mathematics;
- mathematical definition of demographical units accounting their lower (natural) and higher (engineering and knowledge) strata;

The paper is aimed to the carrying out of practical tasks related to the demographical strata in constructing the new world order.

2. THE MAIN FACTS OF HIERARCHICAL MATHEMATICS

Hierarchical mathematics (aed theory) is based on the main law of hierarchical time&space: all strata arise in multiplying and uniting of units of lower strata, create highest strata, and are changed by their sway.

Nowadays the original state of aed A^{λ} has two basic records $^{\times}\alpha^{\lambda}$ and $^{+}\alpha^{\lambda}$ which allow to realize the law defined above. The record $^{\times}\alpha^{\lambda}$ aggregates the continuous theories (and extends their abilities) and the record $^{+}\alpha^{\lambda}$ is more near to discrete ones. A^{λ} in the code of $^{\times}\alpha^{\lambda}$ is defined as follows.

$^{\times}\alpha^{\lambda}$:

$$A^{\lambda} \overset{\gamma}{\underset{\rho}{\longleftrightarrow}} \left\{ {}^{\beta}_{\omega} \overset{\gamma}{\underset{\rho}{A}} {}^{\lambda}_{\sigma} \right\}^{\lambda \xrightarrow[\rho]{A^{\lambda}} \beta}$$

where Λ,λ-level (hierarchical number, time&space), Γ,γ-statute (law, connection), P,ρ- act (process), Ω, ω - unit (state) Σ,σ - construction (contents), B, β - hierarchical haze (new time, arising level), A, α - sway (coordinator).

The details (directions, strata) $\Lambda,\Gamma,B,P,\Sigma,\Omega$ are described by aed original state (by the multiplying of A^{λ} record in its own construction). That is they have all details and abilities of A^{λ} and at the same time they define their individual characteristics.

For instance, in the record of hierarchical haze β^{λ} its coordinator is defined by its own symbol B unlike the original (key) unit A^{λ}:

$$B^{\lambda} \overset{\gamma}{\underset{\rho}{\longleftrightarrow}} \left\{ {}^{\beta}_{\omega} \overset{\gamma}{\underset{\rho}{B}} {}^{\lambda}_{\sigma} \right\}^{\lambda \xrightarrow[\rho]{B^{\lambda}} \beta}$$

The uniting of similar records of all aed details leads to the constructing of highest level unit A^{β}.

$$A^{\beta} \overset{\gamma}{\underset{\rho}{\longleftrightarrow}} \left\{ {}^{?}_{\omega} \overset{\gamma}{\underset{\rho}{A}} {}^{\beta}_{\sigma} \right\}^{\beta \xrightarrow[\rho]{A^{\beta}} ?}$$

In this way the original statute of aed A^{λ} is well defined as well as the images of its ordinary states ($\Lambda,\Gamma,B,P,\Sigma,\Omega$), but the statute of its leading unit A^{β} contains in its construction the hazy (uncertain, vague) symbols. It means that the new time β cannot have the exact definition at time λ. At that time its statute is chimerical and it will be changed in line with the main law of hierarchical time&space.

The record $^{+}\alpha^{\lambda}$ (Fig.1) contains the records of $^{\times}\alpha^{\lambda}$, but arranges them in graphical construction convenient in applied tasks.

In Fig 1: A^{λ} - original state of aed ($A^{\lambda} \leftrightarrow \omega^{\lambda}$), A^{β} - its leading state in time λ; multiplying act ${}^{\times}\rho^{\lambda}$ of original order γ^{λ} (original statute (code, record) of A^{λ}) executes learning process in the units ω^{λ_τ} of construction σ^{λ_τ} which was created in time λ_τ ; thanks to that the lower strata units ω^{λ_τ} turn into the ordinary units ω^{λ}_{τ} in new construction σ^{λ} ; uniting act ${}^{+}\rho^{\lambda}$ in σ^{λ} connects the ordinary units ω^{λ}_{τ} , design and create the new sway $\omega^{\beta} \leftrightarrow A^{\beta}$; in the process ρ^{λ} the image of arising (hazy) time β^{λ} has two strata - ${}^{\times}\beta^{\lambda}$ and ${}^{+}\beta^{\lambda}$ which correspond to the two stages of this process: ${}^{\times}\rho^{\lambda}$ and ${}^{+}\rho^{\lambda}$.

The main records of hierarchical mathematics described above allows one to construct the exact mathematical definition of all strata of real world including demographical ones.

3. HISTORY, CONSTRUCTIONS, ACTIVITY AND GOALS OF DEMOGRAPHICAL UNITS IN HIRARCHICAL SPACE

In order to achieve the aim of the paper it is necessary to take into account the history of hierarchical space as a whole.

In accordance with hierarchical world-understanding at first the physical units ω_{ι}^{ph} have been created and united then in their leading units ω^{ph}_{10} of higher level. (The symbol 10^{λ} means the whole number of units of any level λ which must be connected to create the unit of higher level. Different levels have different numbers of ordinary units in the constructions of their sway). In physical space the leading units are the stars and stars unions (congestions). The stars $\omega^{ph}_{10} \leftrightarrow \omega^{ch}_{0}$ by the act ρ^{ph} unite the ordinary physical units ω_{ι}^{ph} (the most significant among them are the hydrogen nuclei, which belong also to the chemistry time beginning), change them and arrange the changed units along the star layers (strata) - from hydrogen to heavy nuclei in agreement with stars' laws $\gamma^{ph}_{10} \leftrightarrow \gamma^{ch}_{0}$. Then, every chemical nucleus is a symbol of certain star layer - symbolic realization of star history till this layer arising. At the same time it is the symbol of future (chemical) strategy - chemical activity in reactions. This activity in stars is defined only as outlook in the level space (as arising time signs) at the area of physical haze β^{ph}. Star construction is created by astrophysical technology, then - by astrochemical one which unites and changes diverse astrophysical acts.

Astrochemical technology ${}^{\times}\rho^{ch}$ leads certain stars to the bangs: they are divided (multiplied) as original units of chemistry time. Star bang leads to its uniting history renovating in new - astrogeochemical - technology ${}^{+}\rho^{ch}$. This restoring creates several units of chemical time - large scale chemical reactions which contain all symbols of star layers and connect them in accordance with chemical laws γ^{ch} owing to the chemical activity realization as new sway ω^{ch}_{10}. Various states of chemical reactions (gaseous, liquid, solid) are the layers of geochemical construction σ^{ch} and these layers are connected by biochemical units $\omega^{ch}_{10} \leftrightarrow \omega^{b}_{0}$ - leading units of chemistry time and original states of biology level. They are constructed in chemical haze β^{ch}.

Biochemical units (later - cells) ω^{b}_{ι} posses the signs of all states of chemical reactions and unite, change and control them in agreement with biological laws γ^{b} by biological acts ${}^{+}\rho^{b}, {}^{\times}\rho^{b}$. They are multiplied as level initial state by biological technology ${}^{\times}\rho^{b}$, and uniting ${}^{+}\rho^{b}$ of them leads to the arising of multicell units ω^{b}_{10} (the symbols of cell space with all its layers). Biological units change all physical and chemical processes in their constructions σ^{b}; multicell units $\omega^{b}_{10} \leftrightarrow \omega^{d}_{0}$ link cells states and change them: the cells in their constructions are similar with the multicell units and single cells are able to restore the whole multicell unit - the leading state of biology level and the initial one in demographic time which exists in biology level as its hazy stratum β^{b}.

Multicell units ω^{d}_{ι} multiplying ${}^{\times}\rho^{d}$ and uniting ${}^{+}\rho^{d}$ acts in demographical space create the constructions σ^{d} with key units $\omega^{d}_{10} \leftrightarrow \omega^{e}_{0}$ connected by engineering (in future) interactions γ^{d}. They unite and change all times created before in their technologies. The units ω^{e}_{0} are the leading instances at demographic level and the initial states of engineering one in hazy area β^{d} of demographical time.

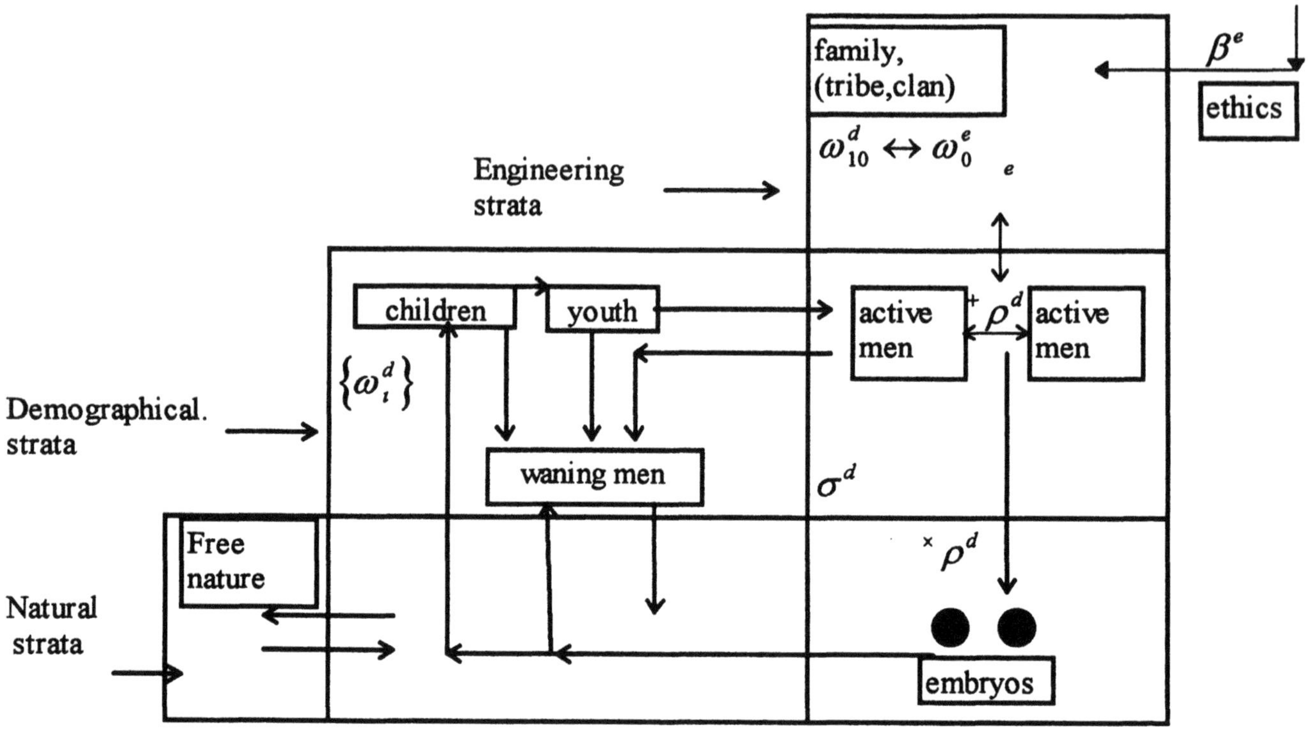

Fig.2 Demographical and engineering strata in time of knowledge strata arising

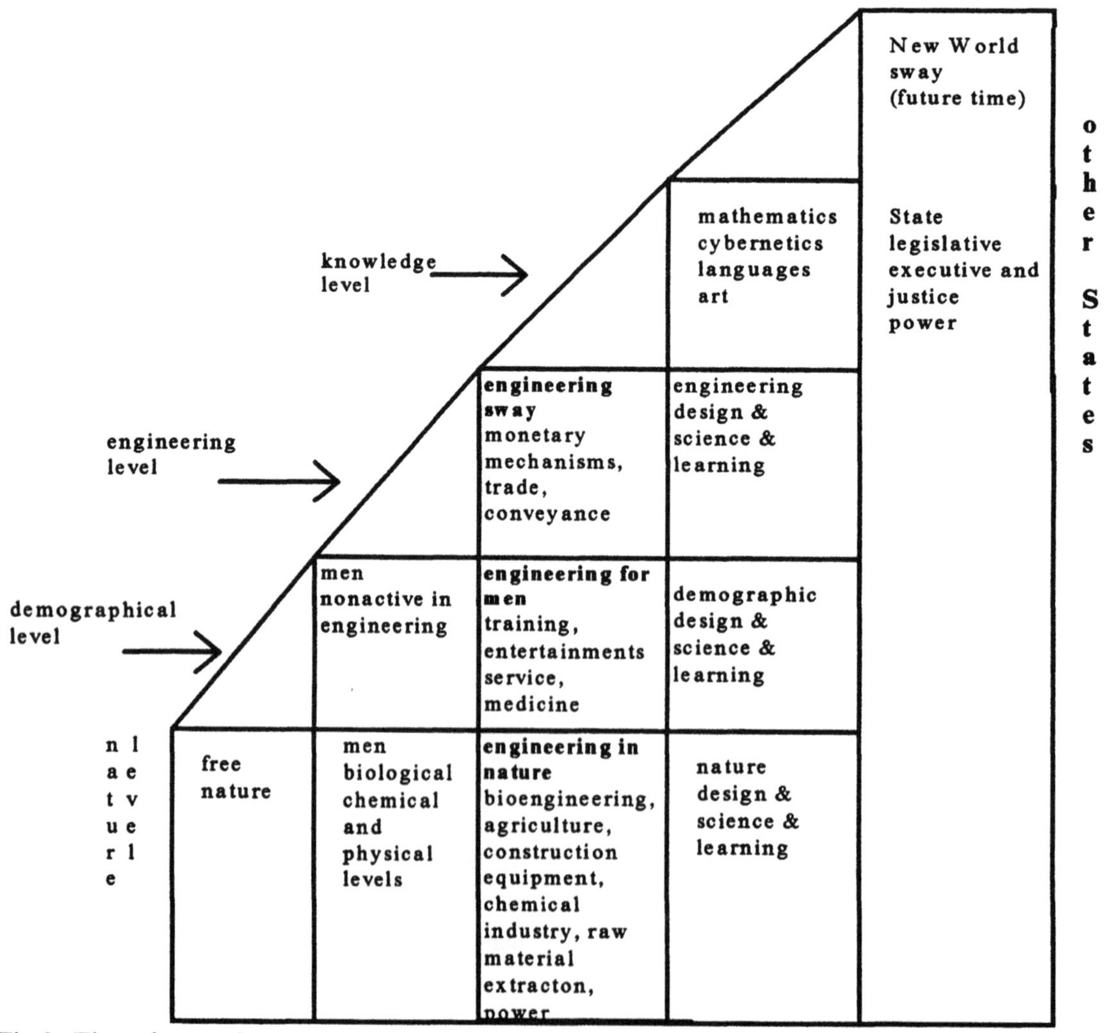

Fig.3. The scheme of sway networks of States; the scheme of date base of demographical strata is included in this network

The units ω_t^e of engineering time connect and change all symbols of lower times. Demographic units ω_t^d in engineering constructions σ^e are able to cohered actions with aims β^e settled before due to the creation of exact symbol images of concrete world strata with their future states. Multiplied by their own technology ${}^{\times}\rho^e$ the engineering units are connected by the knowledge means (language, art and others). The leading engineering units $\omega_{10}^e \leftrightarrow \omega_0^{kn}$ (which connect various engineering technologies) create their statutes - at first in the works of art in engineering haze β^e. The statutes γ^e define their world outlook: their history, current laws of their constructions, the laws of interactions with their environment (other units of their level) and future strategies. The arising of the statutes γ_{10}^e is the beginning of knowledge time. The leading engineering units $\omega_{10}^e \leftrightarrow \omega_0^{kn}$ are the initial states in knowledge space defined in engineering strata as hazy area β^e.

The initial units ω_0^{kn} of knowledge time - the States - create the State power institutes (legislative, executive and justice power), which maintain their statutes γ^{kn}. They include all lower strata (nature, demographic and engineering layers) and the strata of outlook in the level space (the haze β^{kn} of knowledge time): learning, science, art, design. Now the States multiply and unite with new time creation in hazy strata β^{kn}. Any State as well as every unit is the symbol of both its history (creating technology) and its future strategy in arising sway of the world.

Demographical space during knowledge level arising is defined at Fig.2 which is concrete record of Fig.1. Fig.2 contains three levels: natural, demographical and engineering. Natural units (physical, chemical and biological) are included in the construction of demographical units as their own details. The interactions of demographical units are realized, at first, by the exchange of named details. The sway of this exchange is realized by the families (tribes, clans) — the leading units of demographical level (the initial instances of engineering time). The demographical space contains the waning systems (unlucky instances) which cannot work as active systems in engineering ranges. They arise and disappear at the boundaries of demographical level. However other demographical units are able to get sway ranges which contain active systems able to create the families. They multiply, then (being old man or units going out) return to natural strata. The acts of uniting and multiplying are carried out in engineering ranges. This activity attracts active demographical units, such behavior is included into the core of their aims. The space of aims (hazy area) begin to exist in the constructions of demographical units as the most significant range. Together with natural strata the constructions of demographical units contain the images of higher strata, their goals. And these images turn into the concrete organs in mechanisms of higher strata realization (the mechanisms of demographical units maintaining and renovating).

In this way when higher strata arises the new ranges in demographical space are created, and these strata are defined in the constructions of demographical units as new details.

The hierarchical space as a whole in its existent state is imaged by Fig.3.

All times (levels) on Fig.3 are described by strata of their constructions with their sway institutes and their outlooks in the level space which contain hierarchical haze in their construction. The history of every level can be taken into account by considering the current state of highest (knowledge) level. Now it is the construction which connects diverse knowledge units by the field of arising time (thought strata). The construction of knowledge level as a whole contains all other times and is contained in them - its image intersects all time images of level space (engineering, demographic and nature): knowledge is power of all lower strata and has direct contacts with them. The knowledge units (States) are linked by the arising time unit. This single unit is the outlook of them in the level space, the strategy for the future. The unit of thought time is described by the chimeric image (only by the signs of its own construction; its exact images in its environment and the strata of its own sway remain hazy) while the units of lower times are described also by their interactions with the units of their time, by higher laws, and by the highest (aed) statute.

Fig.3. gives a chance to connect and to order the internal and external characteristics of demographical units, to construct their mathematical definition good in the carrying out of the main cybernetic tasks (design&control& learning) at demographical level. It allows to see the significance of engineering and knowledge strata in these tasks. Being the sway of demographical level the engineering and knowledge strata define the conditions of life and work of demographical units, including their aims. The field of aims is defined by Fig.3. They are attained thanks to the intention to move to sway ranges by the acquiring of property at engineering level, professional skills, knowledge, ability for design (art creativity). The sway strata must have the means of this movement maintenance (health protection, entertainment, justice and other). At the same time the errors in engineering and knowledge strata activity are now the main causes of demographical disasters. International and civil wars, bad schools which cannot give knowledge required by real life, unemployment, professional diseases, ecological systems contamination by industry destroy the demographical level. Since that, in order to improve the desired state of this level, it is necessary to change its higher strata.

4. CONCLUSION

The definition of demographical strata (in symbols $A^{\lambda}, \Lambda, \Gamma, \Omega, \Sigma, B, A^{\beta}$ of hierarchical mathematics) allows one to connect the constructions of demographical systems with their activity as units in their environment and with their sway (higher) strata. Unlike other definitions (constructed within the sets' theory framework) the proposed one meets all known requirements of design&control&learning tasks.

Owing to that the new definition is convenient in the elaboration of changeable data bases for demographical layers' describing and these data bases are cohered with sway networks of States and unions of States which arise now throughout the world. Besides, it allows to design the strategy for future investigations of demographical strata. This strategy contains the process of connection of internal characteristics of demographical units (including psychological signs, health and other) with characteristics of higher strata, firstly with that of the areas of creative work — ethics, design, science and learning.

REFERENCES

Caring for the Future. (1996). Report of the Independed Commission on Population and Quality Life, Oxford New York.

Groumpos, P., S. Novikava, et al (1997). Design&Creating of New Engineering Units in Reconstruction Regions. *Proceedings of the Annual Conference of ICIMS-NOE (E.P.9251) on ASI'97 in Life Cycle Approaches to Production Systems: Management, Control, Supervision.* Budapest, Hungary.

Novikava, S., S.Gancharova, A.Zhybul et al (1997). The Statute of Hierarchical Mathematics and Its Cybernetical Maintenance. *Preprints of Eleventh International Conference on Mathematical and Computer Modelling and Scientific Computing,* p.149. Washington, DC, USA.

Novikava, S., et al (1995). State Design: New Way in Exact Sciences. *Preprints of the IFAC Conference on Supplementary Ways for Improving International Stability* - SWIIS'95, p.175-181. Vienna, Austria,.

ART IN STATE CONSTRUCTING PROCESS

A.Burawkin*, S.Gancharova, A.Danichaw*, S.Ioska**, A.Turovich**, I.Padyninogina**, A.Zezula*, W.Zianevich*, A. Zhybul**, Y.Bludaw**, H.Osipava**, Y.Terahaw**, T.Areshka*, A.Astrowskaya**, D.Sarokina**, A.Bludaw**, I.Yuhnevich***

**) Belarusan University of Culture*
***) Hierarchical Multilevel Systems Laboratory,*
Institute of Mathematics&Cybernetics,
220064, Minsk, Belarus, P.O. Box 48, e-mail:imc@org.by

Abstract: The article deals with creative work (art) in context of process of State constructing. Art systems are connected with all strata(levels) of States: natural, demographical, engineering and knowledge. Knowledge stratum (which includes creative work) must coordinate all others. Since that it must define both every stratum as individual level and the whole hierarchical space of known levels. The abstract definitions are constructed in the article on the base of hierarchical images of States and unions of States. Then the theoretical descriptions are applied to concrete instance of Chernobyl disaster. This applying allows to understand the extreme case of instability when the errors of coordinating systems lead, at first, to all levels destroying, then — to the arising of strata able to construct better order in State.

Keywords: art, State constructing, knowledge networks, hierarchical mathematics

1. INTRODUCTION

The activity of art systems (creative work) cannot be accounted and prophesied by existent science. At the same time these systems, being the main components of national consciousness, have great significance in the constructing of new world order (Kile F., 1995). Moreover it is may be that art, which always discovers the new horizons in the world, can allow to see the main lines of the future world more clear than existent science does, and, owing to that, can suggest the most reasonable strategies of desired world attaining.

It is the fact that there are strong links of the process of new States constructing with the strengthening of their national reasons, including art. The professions of many first leaders of States which were constructed or restored in post-soviet space (Armenia, Lithuania, Georgia, Belarus, and many others) are connected with creative work. It attracts to this process the best national forces and allows to understand the internal characteristics of any State and its way in the world.

Since that art belongs to the most significant directions of the activity in States own constructions and in their interactions. However this direction is a zone of high hazard. The lack of its understanding can lead to the errors in internal and external activity of States. (Because of that the former leaders of the countries named above were replaced by the ones which pursue more conservative policy.)The danger of disasters in this activity may be increased when destroyed consciousness uses the new means of cybernetics, (at first, knowledge networks) in the

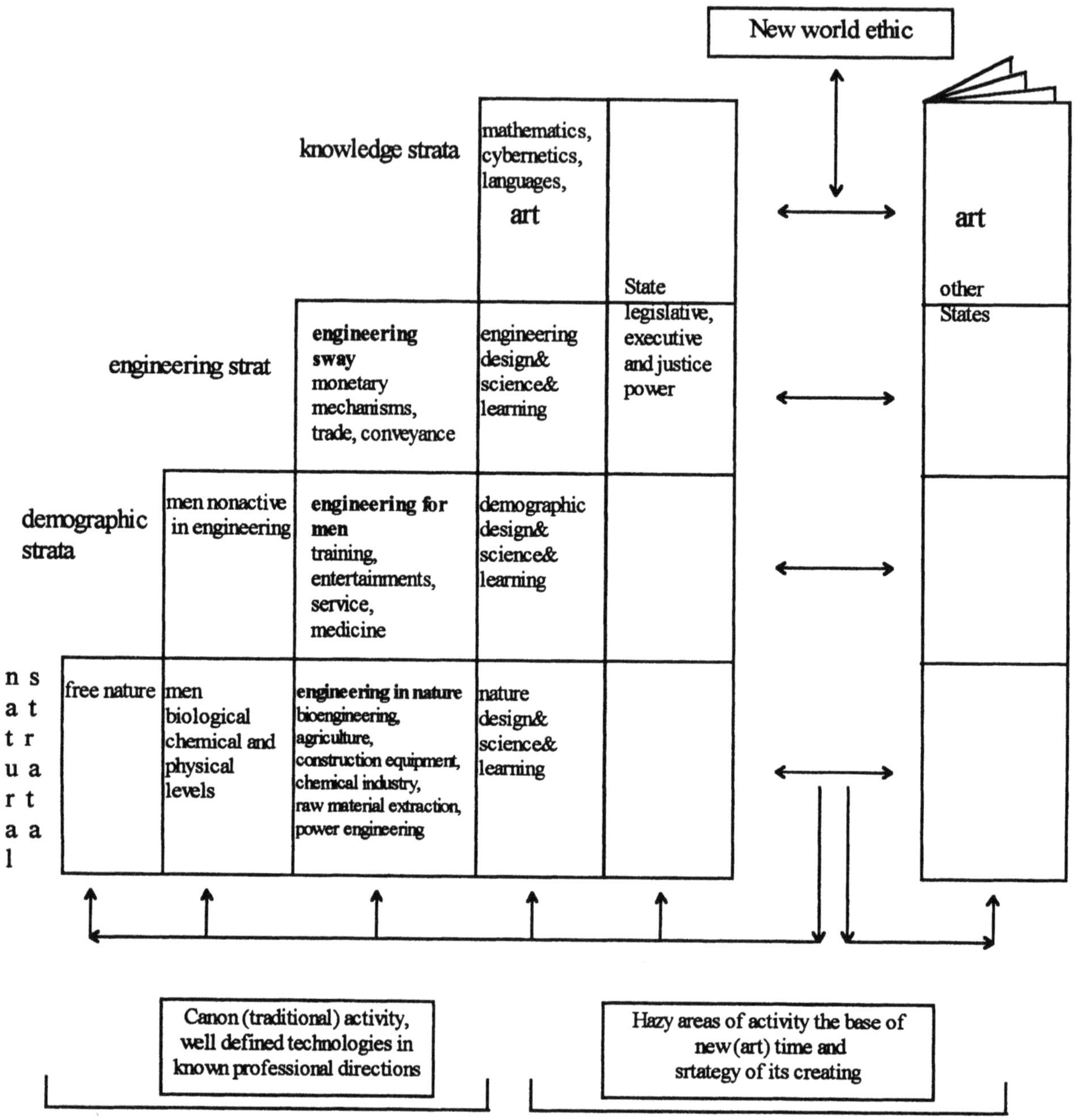

Fig. 1. Art in hierarchical coordinates. Art systems are defined in hierarchical space as active units in the highest current level (stratum). All lower strata (natural, demographical and engineering), whose space is the base of art systems, are now coordinated both by well defined technologies (professional directions created before in hazy areas of art) and by new strategies which arise in the interactions of diverse national reasons. In this way the lower strata are the constructions of art systems, and in their turn the art systems are the sway units which realize all kinds of creative work (design&science&learning) in lower strata. The environment of any art system is the range of other systems of its own level. All art systems have equal rights in this range, but they have individual lines. Their interactions (by exchanging their individual details) allow to construct their arising sway — the new world ethic. When its chimerical definition will be turned into well defined statute, the new stratum in the world will be constructed. The suggested definition gives a chance to connect the strategy of new world ethic creating with the process of States and States unions constructing. Since this definition has its exact image in hierarchical mathematics, it is cohered with new cybernetic means (knowledge networks) in world space.

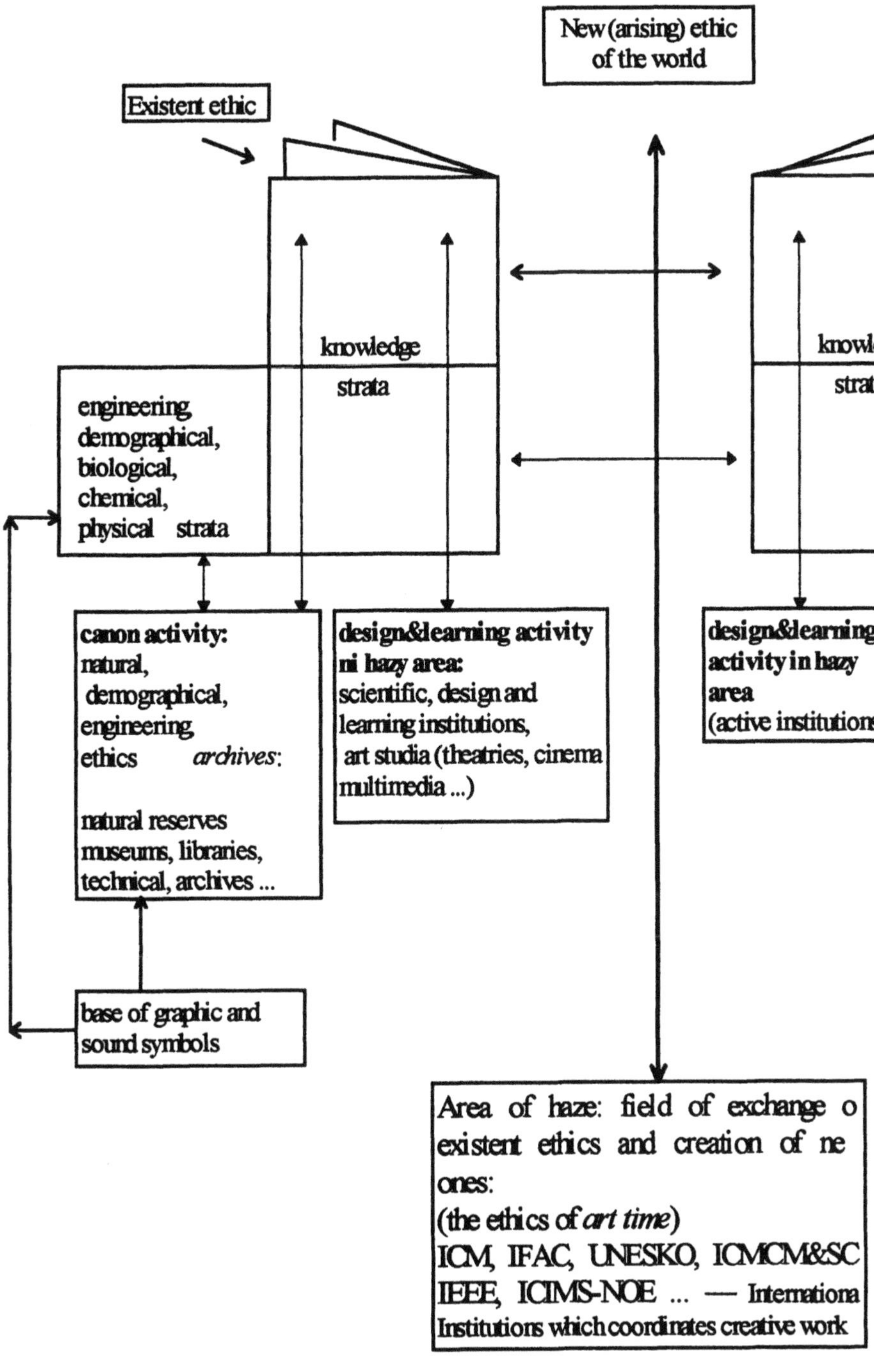

Fig. 2. The area of new world ethics arising. This area includes old (existent) strata (physical, chemical, biological, demographical, engineering and knowledge). Canon (traditional) technologies at first were the directions of creative work in these strata, then they were turned into professional skills. The canon technologies have the archives areas. The archives contain natural reserves, museums, libraries of technical documents, sound and graphical collections and other. The canon ranges of existent ethics are base of its active ranges which contain design, scientific and learning institutions, art studia, and other units of creative work. The new ethic of the world arises in the area of exchange of existent ethics. This area is maintained by all strata created before and it is coordinated by international institutions (ICM, IFAC, UNESKO and the like). The coordinating way of the art area is the strategy of new (art, thought) time constructing, the strategy of the best achievements uniting&multiplying in all world strata.

process of State coordinating. (These networks have now their own negative lines which are accounted in many respects by the weakness of habitual theoretical means which cannot construct the exact definition of the networks, required by the task of their design and control).

In this way the investigations of creative work with taking into account its links with all other directions of activity, mainly with knowledge networks, is now one of the most urgent task. Its urgency was emphasized, for instance, by the conference 'A new Space for Culture and Society — New Ideas in Science and Arts' (Prague, 1996). The works of Don Foresta (Foresta D. et. al, 1995), and Peter Groumpos (Groumpos P., et. al, 1997) connect these tasks with practice of design&control &learning in large scale systems including reconstructing regions.

The given paper describes new method of carrying out of tasks named above. This method is based on hierarchical mathematics, and it is cohered with the key statute of States and States unions constructed with the new mathematical means (Novikava et al, 1991, 1995, 1997). Suggested method gives the new understanding of art, more wide than ones known before. It is convenient in the practice of networking (for example it is applied in the network of the Belarusan University of Culture (BUC) and in the elaboration of data base of art in national program 'Informatization of Culture'. Besides the department 'Information Technologies in Culture' of BUC uses it in learning process. This method is considered below both in abstract and concrete description.

2. THE COORDINATES OF ART IN HIERARCHICAL SPACE OF STATES AND THEIR UNIONS

Hierarchical mathematics (aed theory) allows to connect for any level system its image as active unit in its environment (in the field of other units of its own level), its construction (the units of lower levels with their interactions) and higher levels (its holding systems able to coordinate its activity in the environment).

In agreement with the main law of aed theory the system of any level arises in the space of lower levels whose units create, at first, its chimerical statute, then its aggregated laws. The new level unit must carry out the sway (design&control&learning) tasks in the space of lower strata. In multiplying process the statute of higher level is embodied in the constructions of coordinated units of lower strata. Thanks to that they are learned and their own level becomes equal to the level of their initial sway (the original unit of arising level). The learned units have both the lines of their multiplied origin and individual ones. It means they must be more fine than their initial sway, and they can interact by exchanging their individual characteristics (can learn one another). Then the field of new level systems can realize their uniting process and design their new sway — the origin of more high level. The processes of multiplying (learning) and uniting (design) have hazy characteristics which cannot be accounted and prophesied by lower strata systems. That it these processes are realized as the strategies of art.

The way of hierarchical space creating and the current state of this space are considered in (Novikava et. al, 1995, 97b). Nowadays the regarded space contains the following strata: natural (physical, chemical, biological), demographical, engineering (industry, service, conveyance, trade, monetary mechanisms) and knowledge (languages, art, design, science, learning).

Fig 1 and Fig.2 allow to define art as the unit of knowledge level. The states space of art systems contains national arts. The national arts are active units with changeable individual constructions defined on all strata of States in hierarchical space. This units and the whole art have two main ranges: canon art and new one.

The canon (traditional) arts were the origins (and remain the bases) of existent ethics, they are the base of national consciousness diversity. Existent individual national minds (diverse world outlooks) are symbolic images of the nations. These symbols connect their history, current states, and future strategies. The symbolic images of nations are embodied in the activity of all lower strata (engineering, demographical and natural) in corresponding States. Owing to that the space of known professional directions is divided into several (national) fields with individual characteristics which allow to create international institutions in order to coordinate the interactions in the area of new world ethic arising.

The new world ethic is constructed by art strategy in the area of hazard. Nowadays there are several international institutes which are considered by Harold Chestnut as the instances of arising level units whose multiplying in diverse regions can lead to the increasing of wellbeing in the regions and in the whole world (Chestnut H., 1995). The list of these units may be continued by the including of European Union network ICIMS-NOE (Groumpos P., 1995). This network connects academic (knowledge) and industrial (engineering) strata of States and unions of States. It is large scale hierarchical system able to carry out the task which is under consideration in this paper. The investments of its authors in named direction is the suggestion to include in constructions of international institutes and networks the strata of national arts in broad sense (canonic and new technologies).

The movement to this aim is realized now in Belarus whose current state has to seek the new ways in State constructing process. The applying of the means of hierarchical mathematics gives a chance to define this state and the strategy of its changing with decreasing of danger of errors.

3. THE INSTANCES OF CREATIVE WORK CONNECTING WITH OTHER STRATA IN EXTREME CASE (CHERNOBYL DISASTER)

After explosion in Unit 4 of Chernobyl Atomic Energy Station on April 26, 1986 the large part of radioactive materials (75%) was released to Belarus. Now its dynamics is one the most clear display of connected structure deformations on all levels - from geophysical to knowledge.

Structure deformations began from including the radioactive materials (chemical elements with unsteady physical structure) in atmosphere. Geophysical and geochemical effects (climate conditions, gravitation forces on different relieves, dynamics of water, diffusion in diverse soils - forest, meadow, arable) continue the changes of territory chemical structure.

At the same time the deformations spread along the level space. Interlevel (by means of physical and chemical strata) connections of biological systems with radioactive environment cause their radioactive contamination. Biological systems are restored by cells and mass deformations of cell constructions on large territory and long time tell on general state of biology level to a marked degree.

Biological, chemical and physical deformations change demographical and engineering levels: in Belarus above a third of the territory with its forests, lakes, populated areas, agriculture and industrial structures is radioactive contaminated; nearly 2.5 mln, a quarter of population, live under threatening conditions, for hundreds of thousands the evacuation is required. (Cancer diseases after Chernobyl explosion were increased in Belarus in more than 20 times).

Demographical and engineering activity extends the scale of disaster more intensive than other ways.

Unknown before large scale deformations of demographical and ecological (natural) levels, dangerous state of many engineering structures demand considerable changes on knowledge level - new laws and national programs.

In fact the new strategies of engineering, demographical and ecological systems coordination must be created created.

High uncertainty of knowledge hinders to realize a reasonable control of these events. In many cases a data for coordination are away or contradictory. It is constant occurrence when the experts from different branches of science, design and control in natural, technical and organization systems can not understand each other.

The main reason for that is current state of knowledge level. Widespread means of knowledge strata are onelevel and not suited to consider the laws of hierarchical multilevel systems. Knowledge constructions for individual levels (physical, chemical, biological, demographical, engineering) inherit this restriction. Concrete branches, which are settled by the theories with limited outlook in the level space, can not foresee all effects of own level structure changes on other levels and inversely.

For correction of this state it is necessary to create the system of following level. It must see the constructions, dynamics and connections for any level in hierarchical multilevel system and realize its coherent coordination on all levels (Novikava et al, 1991).

In this way the errors in knowledge strata (bad construction of the atomic reactor caused by irresponsible science and design; criminal behavior of State power after explosion) were the main reasons of engineering, demographical and natural strata destroying.

Moreover, many sources of Belarusan national art are now in radioactive zone and their activities die away. Their end leads to psychological stress, feeling of hopeless in population, and criminality increasing — it is natural consequence of such events.

In line with hierarchical world outlook the national art (highest strata) destroying is the most heavy disaster not only in single State but also in the whole world. In this case the world loses its diversity and thereby the law of level incresing is infringed by system which does it. This system must be changed obligatory.

In order to realize the required changes, the new ranges in Belarusan art, science, design and learning strata arise. They are able both to understand the extreme conditions considered above and to act under them with new strategy of State constructing cohered with the best world achievements in this direction.

The new strategy is aimed to the creation of stable independed State by its own forces in close cooperation with the world institutions which coordinate the process of new world order constructing.

4. CONCLUSION

The definition of creative work in hierarchical coordinates allows to see

- its professional contents (reasoned by existent strata of hierarchical space);
- its own construction (with canon and new strategies);
- its significance in State constructing process (where the hazy (art) strata are answerable for the constructing of new ethics and new orders);
- its connections with advanced technologies, including hierarchical knowledge networks of States and unions of States.

This simple definition is wider than ones known before. At the same time it is more convenient in the carrying out of very heavy tasks — States coordinating and networking.

REFERENCES

Kile, F. (1995). Desired peace. *Preprints of the IFAC Conference on Supplementary Ways for Improving International Stability, SWIIS'95.* pp.147-152. Vienna, Austria.

Chestnut, H. (1995). Improving International Stability and Maintaining Peace. *Preprints of the IFAC Conference on Supplementary Ways for Improving International Stability, SWIIS'95.* pp.11-14. Vienna, Austria.

Groumpos, P. (1996). Advanced Controls for Integrated Manufacturing Systems. *Report for ICIMS-NOE Workshop.* Brussels, Nov. 21st.

Groumpos, P., S. Novikava, S. Gancharova, A. Zhybul and K. Miatliuk (1997). Design&Creating of New Engineering Units in Reconstructing Regions. *Preprints of Advanced Summer Institute'97, ICIMS-NOE.* Budapest, Hungary.

Novikava S. et al (1991). Coordination Technology for Ecological, Social and Engineering System. *Proceedings of Eighth International Symposium on Modular Information Computer Systems and Networks, ICS-NET'91.* p. 47-57. Dubna.

A New Space for Culture and Society — New Ideas in Science and Arts, (1996). *The International Conference.* Prague.

Foresta, D., et al (1997). The New Space of Communication, the Interface with Culture and Artistic Creativity, *A Study for the Council of Europe.*

Novikava, S., et. al (1997a). State Design: New Way in Exact Sciences. *Preprints of the IFAC Conference on Supplementary Ways for Improving International Stability - SWIIS'95.* p.175-181. Vienna, Austria.

Novikava, S., S.Gancharova, A.Zhybul, et al (1997b). The Statute of Hierarchical Mathematics and Its Cybernetical Maintenance. *Preprints of Eleventh International Conference on Mathematical and Computer Modelling and Scientific Computing.* p.149. Washington, DC, USA.

Novikava, S., et al (1997c). Hierarchical Mathematics& Cybernetics in Design of Sway Networks of States and States Unions.. *Preprints of International Conference on Transition to Advanced Market Institutions and Economies, TRANSITION'97.* Warsaw.

PRACTICAL INSTANCE OF STATE DESIGN IN RECONSTRUCTING REGION

P. Buka*, A. Burawkin**, W. Nowik***, Yu. Paciupa****, I. Kuzminich****

* *Hierarchical Multilevel Systems Laboratory, **Belarusan University of Culture, ***Belarusan National Academy of Sciences, ****Grodna State University, 220064, P.O.Box 48, Minsk, Belarus, e-mail:imc@org.by, hmsl@buk.lingvo.minsk.by*

Abstract. The paper describes the applying of theory of hierarchical multilevel systems (aed theory, hierarchical mathematics) to the process of State design in reconstructing regions. Aed theory allows to construct the key statute of States and States unions which connects the descriptions of State internal construction (defined in space of natural, demographical, engineering and knowledge levels), State power institutes, and State external activity. The using of the key statute gives a chance to elaborate the program of State reconstructing of concrete country (Belarus) in line with the requirements of State design process. This method of State design can help to avoid in the reconstructing of single State the hard errors, which can turn this State into the source of world instability. The program is cohered with the strategies of existent institutions in world and regional scales (the United Nations, European Union and other). *Copyright © 1998 IFAC*

Keywords: State design, reconstruction region, hierarchical mathematics, cybernetic technologies.

1. INTRODUCTION

Nowadays the process of new States and unions of States arising is realised, mainly, as chaotic one. Political art rather than exact science and design technologies maintains this process.

The most reasonable instances of this process realising are

- the European Union constructing (when diverse countries are united without the lose of signs of States, and new (higher level) unit is created),
- the United Kingdom level increasing (when Scotland and Wales acquire the characteristics of States, and thereby the UK turns into the unit of more high level than State).

However even these processes belong to zone of high hazard, not to mention the events in post-soviet space and in other regions of the world where States constructing is much more heavy task.

All countries have their individual characteristics and the experience of stable States can not be applied in the regions whose history contains long (unlucky) practice of State construction on the base of the world outlook alien for democratic one.

In order to take into account the experience of developed countries and individual conditions of new States, key statute of States and States unions (constructed with aed theory) were suggested (Novikava et. al., 1995). This statute defines

- State construction&activity on all levels: natural, demographic, engineering and knowledge;
- State sway (constructions, activity, interacts and changing way of legislative, executive and justice power);
- State external activity as the unit in the world (its connections with other States on all their levels);

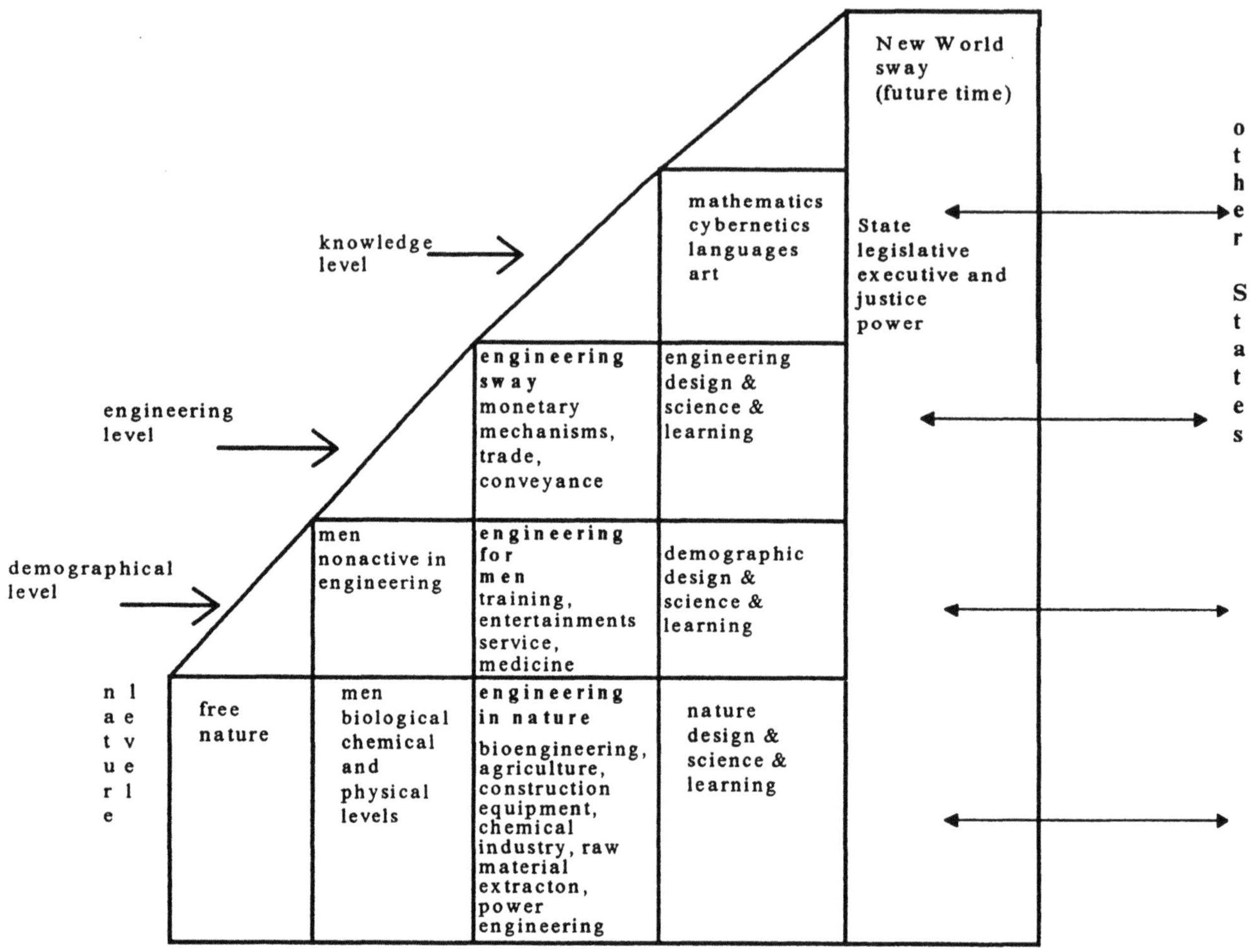

Fig.1 The key statute of States and States unions

- way of statute changing (which also must be cohered with realised turns) with its defence against internal and external illegal acts.

The concrete program for all levels (strata) coordination in State constructing process can be elaborated and cohered by using of this statute (Buka, 1997). The program of State reconstructing includes the required actions on all strata with taking into account their connections.

The paper includes the definition of the general statute of States and States unions, and the concrete program of Belarusan State reconstructing with this statute applying.

2. THEORY

2.1 Key statute for States and States unions

The key statute of States (Fig.1) connects:
the State construction on all known strata (natural, demographic, engineering and knowledge) with taking into account the settlement of lower units by higher strata signs; every layer has direct contacts with all other strata; thanks to these connections it is able to design&control all lower and higher constructions and at the same time it is changed by lower and higher activity; however the connection with higher levels is stronger than the one with all lower statutes: all lower laws are changed by higher strata; nature strata is divided in free and new (caused by higher levels) layers; the demographic layers create the units of engineering and knowledge strata and are changed by them - new layers are created in demographic field by new sway as well as new engineering strata are defined by knowledge units;

the State sway institutions (constructions, activity and interactions of State legislative, executive and justice power); these institutions have concrete realisations for all lower layers of the State; the will and the states of lower levels are strongly connected with the will (outlook in the level space) and with the states of all State sway institutions; these institutions are also connected with arising sway;

the State external connections as the unit in the world (its interactions with other States); the interactions with other States are realised by all State strata: the States exchange the nature, demographic, engineering and knowledge units; this exchange must be controlled by State sway and at the same time it is the reason of interState laws arising; these laws are united in the world statute - the new world sway; (the interactions of State diverse strata are marked on Fig. 1 by the arrows which are under control of the State and world sway institutions);

the way of design&control of the State statute (statute correction and its defence against internal and external illegal actions); this way is defined in the statute image as the technology of joint will of all State strata realisation and as the strategy of statute design&control by arising sway of knowledge level; the strategy of new sway is the elaboration of new knowledge means, the most coherent with the aim of level increasing of all diverse States.

The key Statute of States and States unions allows to determine (in accordance with the aims of State reconstruction: democratisation, the effective economics and so on) the changes of State own construction, State power, and State external connections. The concrete actions, which realised determined changes, are united in the program of State reconstructing.

It is easy to understand that the statutes of all units of lower strata (physical, chemical and other) are the details of key statute. They are described in the space of key statute which connects their own changeable constructions, their interactions with other units of their levels, and their higher units and acts (their sway which is realised in their own constructions in the degree, coherent with their states & acts in the level space).

Thus the programs of reconstruction of concrete units also can be designed. The using of key statute of States and States unions easy allow to cohere this programs and to include them as a detail in the program of State reconstruction.

2.2 The creation of the program of State reconstructing

The facts considered above allow

- to define the requirements to the program of State reconstructing,
- to design the program,
- to suggest a number of concrete actions for State design&control with this program.

The new institutions of State sway and local authority directed on every strata of own construction of the State must be created with the purpose of realising of this reconstruction. Also the new issues (for all directions of State activity) must be introduced in State statute and correspondent parts of legislation must be rebuilt.

Suggested reconstruction of the sway institutions is cohered with the requirements of existent and arising world. Therefore, even if the State reconstruction will be completed, they can be useful until the key statute will be changed.

The main brunch of activity will be the reconstruction of knowledge (thought) ranges and creative activity, which is the key for other levels reconstruction. Also it is need to pay special attention to external links, because of their ability to effect on any State activity.

The program of State reconstruction includes the named and other actions directed on the reconstruction of concrete strata. If it is necessary to consider the individual stratum, it can be described in details: with the definition of changing way of its construction, units, acts, laws, and sway.

Thanks to coherence of the program with cybernetic (design&control&learning) tasks it is convenient in the elaboration of sway network of State.

3. THE PROGRAM OF STATE RECONSTRUCTION

The program includes the next connected parts, which describe

- State external connections and activity in the world;
- State sway (legislative, executive and justice power) and local authorities
- strategies of all strata changing in State own construction

The concrete suggestions for reconstruction of all levels of State (natural, demographical, engineering, and knowledge) and its interactions with other States on these levels are defined in the program.

<u>The aims of Belarus in the world</u>

Belarus ought to take an active part in the UN, and other world and regional institutes aimed to defence and restoration of nature, protection of human rights, coordination of States interactions in industry and trade, maintenance of languages, art, science and learning, prevention of armed conflicts.

It need to keep and restore mutually beneficial relations with post-soviet States (The Community of

Independent States (CIS)) and to establish new connections with other States, with the aims to balance interactions with some States, to diversify them, to protect Belarus against outside harms in times of possible distortions of other States. The advantageous geopolitical state of Belarus should be used for organising of external connections.

It's necessary to pay special attention to interactions in knowledge ranges and creative activity.

<u>**The upgrading of Belarus own construction**</u>

On the level of legislative, executive, and justice power

There are a great need in the upgrading of Belarus Constitution by introducing new issues which must include aggregated exact image of natural, engineering, and knowledge legislation), which must be connected with the existent issues for demographical level.

Executive and justice power should be changed according to the new Statute issues: i.e. to create the executive and justice institutes assigned for State and local authorities in all issues of the Main Law.

It is necessary to create State and municipal networks directed to supporting of power institutes (including institutes of security and protection of State boundaries) by exact knowledge about all strata states&actions, and State external connections.

If the proposed issues are not included in Constitution, the particular laws become incoherent with each other and practically unusable. Such strategic laws as laws directed to protection of nature; using resources; creating and strengthening of private property in trade and engineering; maintenance of science, learning and art; State defence, security and protection of State boundaries; and so on are uncontrolled by Constitutional Court. As the result, the possibility of unpunished abuse of power is created. A people, as well as units of other levels, poorly realises its rights and duties, they can not resist illegal actions of State institutes, they have not appropriate defence against criminal trade and other criminal engineering constructions or natural calamities, and suffer hard damages.

The suggested changes were discussed on all levels of science. They are cohered with the aims of the UNO, other world and European institutes, CIS (Community of independent States), and developed in all concrete directions of State activity.

On the knowledge (creative) level

It's necessary to upgrade the laws, executive and justice institutes directed to design&control of science, learning (including press, radio, TV), art and design of Belarus.

The knowledge level activity, which can and ought to be the main source of State prosperity, continues to be considered as the secondary sphere, and it is mainly neglected by engineering and authority ranges. The scientific results, which are necessary for sensible elaboration of laws, creation of competeable industry and profitable trade, learning of active competent and responsible professionals, protection and restoring of nature, have no wide support. On the other hand the science and design are not responsible for the harms caused by application of destructive theories and technologies, and their own inactivity in State constructing. The learning is not responsible for unemployment, criminality, and growing of sickness rate, caused by the lacks of the necessary level of knowledge and upbringing. The art has no connections with building, constructing of technical wares, conditions of life and work.

Science, learning, art and design in Belarus must get their laws and institutes, which should on the one hand interest industry and trade ranges in investing in knowledge, on the other hand raise the responsibility of creative ranges for State own construction and its way in the world.

The first measures are to be the following: to create the network of learning&science institutes with the aim of studying of Belarusan laws, foreign languages and construction of foreign States, supporting in creation own industry and trade units.

On the level of industry, trade and monetary mechanisms

Engineering legislation must be reconstructed. The main rights and responsibilities of industry and trade units ought to be fixed in Constitution. It's necessary to define the role of industry and trade as for the connection of Belarus with the world, the interactions of industry and trade with creative ranges, population, and nature. The executive and justice power must be changed according the new legislation. There is need in the creating of the knowledge (computer) networks aimed at the swaying of engineering level.

Today Belarus (and other arising States) own science, learning, art, and design don't support its industry, trade, and monetary system. Many creative institutes are outsiders in the process of State reconstruction. In these conditions, the market mechanisms introducing leads in many cases to the destroying of both State engineering and arising private industry and trade, including agricultural. The groundless adopting of laws and institutes of other States, created in other conditions, causes the sharp growth of criminality in industrial and trade ranges, losing of living standards, destruction of nature, and, as the result, it leads to giving the market up.

The rights and responses of industrial and trade units of Belarus, both State and private, cohered with the own construction of Belarus, must be exactly described in Constitution. Nobody can hinder their good activities, prohibit it (it may be done the only by law and according the Constitution), and establish groundless taxes. At the same time engineering units must be responsible for harms of nature, people, industrial and trade units, Belarusan State and its state in the World.

The State must support the applying of new technologies aimed to creation of competeable wares and improvement of living conditions in industry, service and trade. The State ought to support also the activity of engineering units in fields of nature restoring, men health improving, providing of unemployed, knowledge activity, creating and strengthening of links with other States.

First actions are

to create knowledge networks, and concern all data about the private and State enterprises, which may be necessary for checking of their state, their activity directed to nature and population, their interactions with other enterprises in Belarus and other countries, creative abilities, and relations with the State sway institutes.

It's a necessity to carry out denationalisation of unprofitable enterprises and put this process under the control of all interested persons and institutes.

It's needed to maintain actively the process of the creating of the small and middle enterprises (agricultural being among them) to maintain the arranging of their connections with Belarusan and foreign enterprises, to support it by juridical, scientific and learning institutes.

The conditions for foreign investments should be created.

The States own monetary system should be consolidated.

On the demographical level

The demographic legislation must be improved, mainly by indirect means: on the one hand by fixing in Constitution the rights of a demographical units as a creators and owners of engineering and knowledge units, and, on the other hand, fixing the rights to have a good natural, living and working conditions. It is necessary to consolidate existing and create new mechanisms of executive and justice power, which must be responsible for realising of human rights and increasing of the life quality of population. It needs to create knowledge networks with the aim to make exact image of the state of demographical level, the ways of its changing, and render the controllable support of demographical units.

The natural increment of population decreases, but infant mortality, the sickness rate of serious illness, and child criminality increase. The youth is not ready for self-dependent existence and work. The great part of able-bodies population is not able to keep their families. The rate of unemployment and hidden unemployment grows (mainly among the youth and women). Also the number of hard crimes and suicides increases. The elderly persons and invalids, whom form about the one third of population, practically are over the limits of survived.

Hard state of demographical level is caused by the state of other levels of Belarus: nature pollution as the result of Chernobyl disaster and other technogen influences; imperfect medical service; poor leaving and work conditions; the impossibility to get the knowledge, moral and professional training, which need for getting a good job in modern conditions, in existent learning institutes; the lacks in executing institutes directed to distribution of means, and control of criminality (including the crimes in the privatisation of State property); incoherent state of legislation, which has no science&learning support, and do not correspond with the local conditions; lack of comprehension of cardinal changes in other post-socialist States, impossibility to keep former connections, which was counted on State property in sway ranges, insufficient care to establishing of new industrial, trade, knowledge connections with this countries and other countries of the world.

First actions for the demographical level ordering must be mainly directed to other ranges of the State: the learning and training system, engineering (including dwelling constructing), medical service, natural conditions. The main measure, required for control of living, learning and working conditions, distribution of State resources, is the creating of knowledge networks (based on the new cybernetic), designed for the carry out of the tasks of executing, justice and legislative power and local authority, and for support of participation of the active populations in control of demographical and other ranges.

On nature level

Nature legislation ought to be created with the purpose of protecting and restoring of nature, ensuring of reasonable using of the soil and bowels of the earth, water and air, flora and fauna, and fixed in the Main Law of Republic of Belarus. Connections of this legislation with demographical, engineering, and knowledge ones must be established, with the aim to activate the protection and renovation of nature and to increase the responsibility for the state of nature. It's needed to upgrade existent, and create the new executive and justice institutes corresponding to the changes in nature legislation. The knowledge networks with the concrete data about the state and changes in nature should be created on the base of new knowledge technologies. It must be meant for using by all interested persons and institutes as well as State and local authorities.

In the result of Chernobyl disaster and other technogenic, scientific, design and authority criminals, (including the unreasonable irrigation, forest destroying, and pollution of soil, water and air) the nature of Belarus is in critical state. This state creates the serious menace for the population life and health, the engineering and service extending (including the agriculture, recreation, and tourism industry), activity of upbringing and learning institutes. The world unique sources of ethnic arts in many places were destroyed. The State and local authorities must have great expenditures for restoring of nature (and for manufacturing in unsuitable conditions).

The first actions are to be the next: to create sway networks for control and changes of nature, using the space satellites, air photography, and land observation stations; to adopt "green" and resource-saving technologies in engineering.

Resume

This program is the only one, which includes and connects all tasks of State design, creates the possibility of the coordinated arranging of all levels of the State and its interactions with other States. It is impossible to realise a wide range of tasks: to create the civil democratic State, to restore nature, to grow the well-being of population, to increase the level of industry, science, learning and art of Belarus without realising the defined actions of the suggested program.

4. CONCLUSIONS

Nowadays the dividing of diverse States (Soviet Union, Yugoslavia) allows to begin the process of democratisation in new States. But this States remain internally unstable and can excite external conflicts, which may lead to a local and global wars.

One of the causes of instability in new States is the habitual strategy of State design, when the laws of known States are realised as the means for new State settlement. Such legislation is well cohered with real conditions of their States-sources, but it can not describe correctly the state of concrete strata and their interactions in new States, and has no means for detecting their own contradictions. As the result the discrepancies in the laws with concrete construction of new States are discovered later, when the serious distortions take place.

Therefore the new strategy of State design (the strategy of sensible creating and changing of States and their unions) must be created now for reconstructing regions. It is defined in the paper on the base of general State statute, which has identical understanding in all languages and takes into account all diversity of States own constructions&activities.

This strategy may be turned into the program of any political party in reconstructing country if to divide the aims of concrete strata (which this party mainly present: nature (ecologists), demographic and labour (social democrats), and so on). The fulfilment of the program of State reconstruction creates the way and mechanisms for its future amending.

REFERENCES

S.Novikava, S.Gancharova, S.Daronin, A.Burawkin, P.Buka, et. al. State Design: New Way in Exact Sciences. *Preprints of the IFAC Conference on Supplementary Ways for Improving International Stability - SWIIS'95.* p.175-181. Vienna, Austria, 1995,

P.Buka. Practical Using of Systems Theory in State Constructing. *Preprints of International Conference on Transition to Advanced Market Institutions and Economies, TRANSITION'97.* p.72-73. Warsaw, 1997

EAST-WEST ACADEMIC COOPERATION WITHIN THE FRAMEWORK OF TEMPUS-PHARE PROGRAM

O. Pastravanu, M. Voicu

Department of Automatic Control and Industrial Informatics
"Gh. Asachi" Technical University of Iasi
Str. Horia 7-9, 6600 Iasi, Romania
Tel: +40-32-116502, Fax: +40-32-214290, email: opastrav@ac.tuiasi.ro

Abstract: TEMPUS forms part of the overall program PHARE initiated by the European Union for the economic and social restructuring of the countries of Central and Eastern Europe. It has brought a historic contribution to University development in these countries, which represents a large-scale phenomenon with a considerable impact on the whole system of institutions and individuals engaged in Higher Education activity. The TEMPUS projects including Romanian partners support the national strategy promoted by the Ministry of Education for restructuring of the Higher Education in the spirit of internationalization and full academic recognition. The contribution to University development in the Central and Eastern European countries highlight the importance of pre-accession strategies to the European Union, implemented by TEMPUS-PHARE, as well as the future potential offered by SOCRATES-ERASMUS and LEONARDO DA VINCI programs. *Copyright © 1998 IFAC*

Keywords: International stability, East/West/North/South relationships, Educational and behavioral aspects, Sustainable development, Global development: regional impact

1. INTRODUCTION

TEMPUS forms part of the overall program PHARE initiated by the European Union (EU) for the *economic and social restructuring* of the countries of Central and Eastern Europe. Within the framework of PHARE, *human resources development* has been identified as one of the priority areas for cooperation between the (EU) and Central and Eastern European countries. Thus, TEMPUS has been designated as a major human resources development scheme in the *Higher Education sector* which aims to strengthen its ability to meet *educational and training needs* (Cerych, 1990), (Teichler, 1990), (Barr, 1994). This scheme was adopted by the Council of Ministers of the European Union in 1990, and extended in 1993 for a second 4-year phase (TEMPUS II 1994-1999).

TEMPUS, like the PHARE program overall, was launched as a fast aid strategy for the new democracies of Central and Eastern Europe, born at the end of the Cold War period. In this respect, it is worth mentioning that the future political and economic alignments in Europe were not predictable at the time of the program's launch. At the meeting of the European Council held in Copenhagen in June 1993, it was agreed that those countries of Central and Eastern Europe signatory to the Europe Agreements of association with the EU should become members of the EU, subject to their wish to do so and to their eventual ability to satisfy the economic and political requirements for membership. Later on, in December 1994, at the meeting of the European Council held in Essen, a *detailed policy* was adopted for *preparing the accession* to the EU of the associated countries of Central and Eastern Europe. The policy *referred explicitly to the role* of the TEMPUS in the restructuring of Higher Education in the associated countries and *formulated the basic requirements* for the further *admission* of these countries in the EU programs SOCRATES-ERASMUS and LEONARDO DA VINCI (Gwyn, 1996).

Because of the major importance of the factors for change, the contribution of TEMPUS to the extended cooperation process was high. Thus, since its launch, the first phase of the Program supported about 750 collaborative projects; in 1994 (the first year of the second phase) 239 new projects got financed (Gwyn, 1996). These projects, almost 1000 in all, covered a very *wide area of development activities*, ranging from the University curriculum to the student mobility and reported impressive achievements, in both extent and quality, for the participating institutions.

This paper aims to create a brief overview of the *large-scale phenomenon* generated by the TEMPUS-PHARE in the whole system of institutions and individuals engaged in Higher Education activity in the Central and Eastern European countries (sections 2 and 3). A particular interest focuses on the specific needs and development of the Romanian Universities (section 4). For a concrete illustration of the *correlation between goals, activities and outcomes* involved in such an international academic cooperation, the example of the COMPANION project (Companion, 1996), implemented *at the departmental level* by six Romanian and four EU Universities, is given (section 5). The links established by the departments within the TEMPUS scheme are encouraged *to be extended at the institutional level*, by bilateral agreements between the Eastern and Western Universities, within the framework of SOCRATES-ERASMUS program (section 6).

The overall organization of this paper, as well as a number of concrete details about the collaborative activities supported by TEMPUS-PHARE reflect the experience accumulated by the Faculty of Automatic Control and Computer Engineering of Iasi, Romania, that has been enrolled as a partner in five TEMPUS projects: HECE (Hece, 1990), IMPACT (Impact, 1991), DISCO (Disco, 1994), CNTP (Cntp, 1995), COMPANION (Companion, 1996). The first two projects listed above brought together partners from several Central and Eastern European countries (Czech Republic, Poland, Hungary, Slovenia and Romania) and got funded during the first phase of TEMPUS. The other three projects were granted by TEMPUS II and involved only Romanian Universities. The CNTP and COMPANION projects are on going, and the management of COMPANION is coordinated by the aforementioned faculty. Moreover the Department of Automatic Control and Industrial Informatics belonging to this faculty has already signed SOCRATES-ERASMUS bilateral agreements in the field of Control Engineering for the academic year 1998/1999 with four EU departments that are also partners in the COMPANION project.

2. OBJECTIVES OF THE TEMPUS SCHEME

The main goal of TEMPUS-PHARE is to ensure the support for the *development and renewal of* Higher Education, relying on the joint activities and relevant mobilities. Within this context, emphasis is placed on *curriculum development, reform of Higher Education structures and skill-related training of students and academic staff.*

The TEMPUS program has exhibited a number of transitional features intimately related to the broader European context within which PHARE operates. Some of these features are at the organizational level and reflect a post-Maastricht step in the recognition of the legal competence of the UE in defining the *European dimension of education* (Gwyn, 1996).

In the first years of operation, TEMPUS functioned essentially as a fast-response system, designed to keep pace with the dramatical changes of 1989. In an attempt to provide a *rapid benefit* to the universities of all the new democracies, the Program was constructed in such a way that the same modalities served the needs of all these countries (Gwyn, 1996).

More recently, the *diversity of needs* for Higher Education has become apparent (Vlasceanu, 1993), (Whitman, 1993). An obvious illustration of this diversity can be found in the thorough difference of activity scale implemented by TEMPUS in the countries of Central and Eastern Europe. Behind these easily drawn contrasts, there lies the far more important fact that each of the countries participating in TEMPUS has its own economic and developmental needs (Teichler, 1990), (Barr, 1994). Such major differences affect the essential context in which TEMPUS is currently operating in each country.

3. JOINT EUROPEAN PROJECTS

Joint European Projects (JEPs) are *consortia of institutions* that cooperate in activities designed to promote the restructuring of higher education in Central and Eastern Europe (Tempus, 1992-1997-a). From the very beginning, JEPs have represented the keystone of the TEMPUS Program and have focused on the following *main objectives*:

- *Curriculum development* and associated activities (such as staff retraining and materials development);
- Review and improvement of *university management*;

• Development of the *structural capacities* able to ensure the collaboration with *industrial enterprises*;

• Creation of *university networks* for the organization of *student mobilities*.

The diversity of needs within the West-East dialogue has been reflected for some time in the *national priorities* which are set annually for TEMPUS JEP applications (Tempus, 1992-1997-b). At the beginning, there was a broad similarity between the developmental trends of the Central and Eastern European countries, but this gave way to the adoption of national priorities which nowadays reflect *the diverging needs of each individual country*. Therefore TEMPUS outcomes have step by step become quite different from one country to another.

More recently, developments at the level of the European Union have given the selection of priorities a new dimension, *by the integration* of TEMPUS into *a very complex strategy* aiming to ensure a harmonious evolution for the future economic and political life in Europe.

4. EVOLUTION OF PRIORITIES FOR ROMANIA

The national contexts within which TEMPUS operates in the Central and Eastern European countries share many features but are distinct in other respects. As regards Romania, it is of crucial importance to understand the *unfortunate effects of the long period of isolation* from the European academic community on the mentalities of University staff involved in changes (Pastravanu and Lazar, 1992). Such changes taking place in a global framework of the change-around of the entire society may easily become traumatic. Some international reports and monographs were even talking about a sort of resistance to major changes, visible especially to the older generation of academics (e.g. Gwyn, 1996), educated under the auspices of authoritarianism and totalitarianism.

Starting with the academic year 1992/1993, the priorities for Romania displayed their own separate characteristics, gradually displacing the center of gravity towards the following *main areas* (Tempus, 1993-1997-b), (Beju, 1993). (Mihailescu and Vlasceanu, 1994):

• Development of short cycle Higher Education courses and their extension into continuing education;

• Development of cooperation between Universities and enterprises via the creation of University Centers which offer training/retraining courses to enterprises;

• Modernization or restructuring of existing long cycle courses to cater for the needs of European integration;

• Preparation for participation in EU inter-University cooperation programs through mobility projects;

• Restructuring of primary and secondary schoolteacher training through curriculum development and the introduction of modern teaching methods.

The Romanian Ministry of Education acknowledges the TEMPUS contribution as a major resource for development in the sense that the *financial investment* has a great value relative to the proportion of the national budget allocated for development purposes. Therefore the impact of this investment on the evolution of Higher Education is of very considerable interest to the Ministry that, in this context, has became responsible (in cooperation with the National TEMPUS Office), for designing the strategy of national priorities with a view to the alignment to the Western European standards.

5. CONCRETE COOPERATION IMPLEMENTED AT THE DEPARTMENTAL LEVEL BY THE PARTNERS OF THE COMPANION CONSORTIUM

The TEMPUS JEP 11467/1996 "COMPANION" represents the result of close links existing between the six Romanian Universities offering complete educational programs in Control Engineering, on one hand, and their fruitful cooperation with four EU Universities during the time interval 1991-1995, on the other hand (Companion, 1996). The *central objective* of this JEP is to develop *an inter-university network* for supporting *student mobility* and *academic recognition* in the field of Control Engineering, with a view to a long-term cooperation at the departmental level between the partners listed below:

• "Gh. Asachi" Technical University of Iasi, Romania (Department of automatic control and industrial informatics) - coordinator;

• Vienna University of Technology, Austria (University Extension Center) - contractor;

• "Politehnica " University of Bucharest, Romania (Department of automation);

• The Technical University of Cluj-Napoca, Romania (Department of automation);

• The University of Craiova, Romania (Department of automatic control);

• "Dunarea de Jos" University of Galati, Romania (Department of automatic control and electronics);

• "Politehnica" University of Timisoara, Romania (Department of automation and industrial informatics);

- Vienna University of Technology, Austria (Institute of handling devices and robotics);
- "Gerhardt-Mercator" University of Duisburg, Germany (Department of automatic control);
- The University of Gent, Belgium (Department of control engineering and automation);
- The University of Sheffield, United Kingdom (Department of automatic control and systems engineering);
- FRONIUS Welding Machines, Austria (Industrial enterprise).

To meet the central objective stated above, the necessity to accomplish the following *basic goals* has been identified:

(Gl) Getting the *compatibility of the educational programs* in Control Engineering offered by the Romanian Universities, with similar programs existing at the EU Universities. Key aspects herein involved are detailed below:

- G1.1. Getting the compatibility of the *curricula* of the Romanian Universities with the curricula in use at the EU Universities based on the selection of a representative set of common courses, for both core and options.
- G1.2. Adaptation of the *syllabuses* used at the Romanian Universities to ensure the compatibility with the EU Universities for the common courses selected at G1.1.
- G1.3. Increasing the role played by the *students' practice in industrial enterprises* in the educational programs offered by the Romanian Universities, based on the experience of the EU Universities in cooperating with industry.

(G2) *Academic recognition* of study periods abroad, based on the compatibility of the educational programs (see G1) and the introduction of the European Credit Transfer System (Ects, 1995) in the Romanian Universities.

(G3) Reinforcement within the Romanian Universities of the *authorized institutional structures* able to give specific assistance in preparing the students for the mobility programs. Emphasis is placed on:

- G3.1. Improving the cooperation between the departments directly involved in the JEP and the International Offices (already existing in the Romanian Universities) with a view to facilitate students' enrollment in the mobility program.
- G3.2. Enhancing the role of the counseling services given to the students applying for mobility grants.
- G3.3. Enlarging the capacity of modern language departments in providing regular and intensive courses of EU languages.

(G4) Creation of an appropriate *communication infrastructure* to permit the long-term cooperation between the Romanian and EU Universities.

The specific ways to achieve these basic goals have been grouped into seven *main activities*, as follows:

Support activities:

(A1) Improvement of the logistics and communication infrastructure within the Romanian Universities involved in the network;

(A2) Students' counseling and preparation for study periods abroad;

(A3) Improvement of the educational programs offered by the Romanian Universities, in order to ensure the compatibility with the EU partners and the academic recognition of the study periods.

Networking activities:

(A4) Study periods and practical placement abroad;

(A5) Monitoring the study periods abroad;

(A6) Intensive courses for postgraduate students;

(A7) Organization and coordination of the university network.

The three support activities (running in parallel) focus on the development of the *institutional structures* and the *communication capabilities* in the beneficiary departments, as well as on the *procedural aspects* required by the internationalization of the educational programs and academic recognition. The successful accomplishment of these activities lays out the background for the networking activities and ensures the *long-term operation of the mobility network*.

The COMPANION project *joins the general restructuring objectives envisaged by the Romanian Ministry of Education* and *fully suits the national TEMPUS priorities*.

It is worth emphasizing that the EU partners acting in the COMPANION consortium have a *large experience in assisting the development of Higher Education* in the field of Control Engineering in Easter and Central Europe (e.g. TEMPUS JEPs 886 HECE (Hece, 1990), 2011 IMPACT, (Impact, 1991), and 7759 MODIFY (with Hungarian partners)), or in Latin America (e.g. ALFA projects 2048(5) RAPII, and 2318(2) FAVI).

The development of the communication capabilities of the Romanian partners (which represents a key point in the success of this JEP) is strongly related to the progress of the project "Romanian Computer Network for Education and Research", implemented with funds from the Romanian Ministry of Research and Technology, as well as by the help of the finances allocated for the TEMPUS JEP 9149 CNTP (Cntp, 1995).

The expected *outcomes* of the COMPANION project refer to the following aspects:

(O1) The Romanian departments will extend their INTERNET *facilities*; this will allow the creation of several *home pages* presenting the profile of the partners and the current evolution of the JEP.

(O2) Each Romanian department will organize tutoring services; main emphasis will be placed on *students' counseling* for their mobility programs.

(O3) More than half of the grant is allocated to support a large number of *student mobilities* (50-60 from Romania to EU and 10-15 from EU to Romania). Besides the direct impact on the students' training, these periods abroad, monitored carefully, will give a *continuos feedback* about the compatibility created between the Romanian and EU programs in Control Engineering.

(O4) About 90 Romanian postgraduate students will attend presentations on *modern topics* of the Control Engineering given by EU speakers, within the framework of three *intensive courses.* A large number of postgraduate students, not only Romanians, will find the lecture notes available on INTERNET, in the JEP home pages.

The partners of the COMPANION consortium have already identified several ways to support the *continuity of this cooperation* after the end of the TEMPUS funding, among which the SOCRATES-ERASMUS program plays a crucial role.

6. CONTINUATION PERSPECTIVES AT THE INSTUTUTIONAL LEVEL OPENED BY SOCRATES-ERASMUS PROGRAM

The links established by the departments involved in the TEMPUS JEPs are encouraged to be extended *at the institutional level,* by *bilateral agreements* between the Eastern and Western Universities, within the framework of SOCRATES-ERASMUS scheme, which is going to become operational in Central and Easter Europe starting with the academic year 1998/1999.

Unlike the reduced number of persons involved at the departmental level in the management of the TEMPUS projects, the preparation and implementation of the SOCRATES-ERASMUS cooperation requires a more complex contribution of *various categories of people* from the participant Universities (Socrates, 1996):

- *teaching staff* directly involved in the academic aspects of cooperation activities (the choice of study abroad programs, curriculum development, academic recognition);
- *deans of faculties* and *heads of departments* (since the real motivation of the cooperation is created at the level of the structures directly responsible for the programs of study);
- *administration staff* at all levels directly supporting academic cooperation activities (student registration, accommodation office, academic records);
- *international relations* officers;
- *students* and *student associations* (that may be very efficient in creating an international climate);
- members of *policy-making bodies* at different levels of competence (academic councils, faculty boards, boards of deans etc.);
- *national authorities* (that are responsible for the coherence and complementarity of the programmed goals).

7. CONCLUDING REMARKS

The TEMPUS Program has brought *a historic contribution* to University development in the countries of Central and Eastern Europe. This contribution represents a large-scale phenomenon with a considerable impact on the whole system of institutions and individuals engaged in Higher Education activity. *Concrete benefits* and *outcomes* would have been quite unimaginable without the support of TEMPUS-PHARE. In this context, it is worth emphasizing that the launch of TEMPUS triggered a *very high rate of response* from the Universities of Western Europe.

As regards the *concrete activity* of the COMPANION consortium, the JEP joins as partners all the Romanian Universities that have a large experience in students training for Engineering, in general, and Industrial Automation, in particular. Since these six universities are *representative institutions* for the overall development of the Higher Education System in Romania, *the creation of the student mobility network* in the field of Control Engineering and the *appropriate dissemination* of the experience gained by the departments directly involved in the JEP are expected to bring substantial contributions, at various levels, to the restructuring process initiated by the Romanian Ministry of Education. In this context, the following levels of impact are envisaged :

- Staff members of the Romanian departments involved in this JEP will be able to keep pace with the future changes in the education programs of the EU Universities.

- Students who participate in the mobility programs will have the experience necessary to apply for postgraduate programs abroad (M. Sc., Ph.D.).

- The Romanian faculties involved in this JEP will extend and update their *official agreements* with the EU faculties concerning the mutual academic recognition (for medium and long term).

- The results of this JEP along with other European programs including Romanian partners will support the *national strategy* promoted by the Ministry of Education for restructuring the Higher Education in the spirit of *internationalization and full academic recognition of the studies* (for long term).

All these contributions to University development in the countries of Central and Eastern Europe highlight the importance of *pre-accession strategies* to the EU, implemented by TEMPUS for education and training, as well as the future potential of SOCRATES-ERASMUS and LEONARDO DA VINCI programs.

REFERENCES

Barr, N. (1994) Labor Markets and Social Policy in Central and Eastern Europe, Oxford University Press.

Beju, I. (1993). Quality assessment and institutional accreditation in the Romanian Higher education system, *Higher Education in Europe*, vol. **xviii**, no 3.

Cerych, L. (1990). Renewal of Central European Higher Education: issues and challenges, *European Journal of Education*, vol. **25**, no 4).

Cntp (1995). TEMPUS S-JEP 9149 "CNTP" - *Computer network training program.*

Companion (1996). TEMPUS M-JEP 11467 "COMPANION" - *EU compatible training in Industrial Automation.*

Disco (1994). TEMPUS S-JEP 7101 "DISCO" - *Development in Romania of short term Higher Education in Computing centered on distributed processing and its applications.*

Ects (1995). *European Credit Transfer System – User's Guide*, Office for Official Publications of the European Communities, Luxembourg.

Gwyn, R. (1996). *Tempus Country Monograph N° 3: Romania*, Office for Official Publications of the European Communities, Luxembourg.

Hece (1990). TEMPUS S-JEP 886 "HECE" - *Higher Education in Control Engineering.*

Impact (1991). TEMPUS S-JEP 2011 "IMPACT" - *Improvement in Automation and Control Technology.*

Mihailescu, I. and L. Vlasceanu (1994). Higher Education structures in Romania, *Higher Education in Europe*, vol. **xix**, no 4.

Pastravanu, O. and C. Lazar (1992). A professionally disillusioned generation - Can there be found any remedies? In: *Cultural Aspects of Automation* (Forslin, J. and P. Kopacek (Eds.)), 77-80, Springer Verlag Wien New York.

Socrates (1996) *Manual of good practice on Erasmus under the Socrates program*, Office for Official Publications of the European Communities, Luxembourg.

Teichler, U. (1990). Higher Education and employment: changing conditions for diversified structures of Higher Education, *Higher Education in Europe*, vol. **xix**, no 4.

Tempus (1991-1997-a). *Guide for applicants - Guidelines for Joint European Projects*, Office for Official Publications of the European Communities, Luxembourg.

Tempus (1993-1997-b). *Guide for applicants - Tempus Priorities*, Office for Official Publications of the European Communities, Luxembourg.

Vlasceanu, L. (1993). Quality assurance: issues and policy implications, *Higher Education in Europe*, vol. **xviii**, no 3.

Whitman, I. (1993). Quality assessment in Higher Education: issues and concerns, *Higher Education in Europe*, vol. **xviii**, no 3.

FOR HIGHER FLEXIBILITY ON THE LABOUR MARKET

Magdalena Kalaidjieva

Institute for Control and System Research, Bulgarian Academy of Sciences
1000 Sofia, PO Box 119, Tel. +3592-732614, Fax +3592-703361
Email: kalaidji@mbox.cit.bg

Abstract: Employment is usually understood as full or part time job. An alternative might be a combination of "small unit of labour" (SUL) with a part time job, applying different training, supplementing each other for a better fulfilment of personality. The idea is suitable for very poor societies with a good education level, e.g. in the post communist states. The author analyses the social and public environment of a citizen as a system interacting with her/him. A SUL may make labour market more flexible and, added to the techniques of the Global Information Society, to help solving some global problems.

Keywords: Units, Agents, Socio-technical system design, Information depth, Optimisation problems, Organisational factors, Random processes, Process models

1. INTRODUCTION

This paper is an attempt to apply systemic view on the labour market in a wider problem field, together with connected persons and institutions, in order to find supplementary possibilities for small size investments, enhancing quality of life or starting activities, which may grow to private enterprises or not, which may increase the private sector of economy and will increase the market for imports and home made goods. Thus, main accent is pointed on using the natural intention or desire to invest, to enhance quality, to create values, "to be useful to those one loves, to people or nature" or just "to do a good thing". The idea is to create a system acknowledging, absorbing and assimilating the slightest hint of intention to apply for work with stimulating readiness. A system, that does not show different attitude due to the amount of investment and invested effort.

This might also have been the initial idea of the "mass privatisation", and which, not only to my opinion, failed in its forms and content, because it started solving the problem "from the wrong end". It started from excusing for old mistakes of control and rulership and compensating it with fictive money. It started from granting to people values, which are actually their own, from "amending broken pieces", from and by sweeping away the traces of theft, plunder, robbery of the values they created, by transferring one type of stolen goods into another to be stolen once more. That is how people feel about it. And they see, that the purpose and the instruments strengthen their suspicion, on the one hand, and do not enhance the economic situation essentially, on the other. As far as I am informed, similar was the situation in all economies in transition. People first asked themselves why they should be fooled, time elapsed, and hunger made them criminals. Reducing crime and corruption in the destabilised Bulgarian society requires not only to bring before court the guilty (what is undertaken during the recent months, at last), but to give people opportunities to work or organise workers for their existence.

Problems of social system design have been keeping the attention of several generations with Mary P. Follett, (1918), Jan Smuts (1926), Martin Buber (1958), A. Ronald (1986), Frank Michelman (1988), R. A. Dahl (1989), R.N. Bellah, et al (1991) to mention only few of them. The need of conscious creativity in social design grew fast during the recent years with books and authors like B.H. Banathy (1996, 1989b), D.E. Eberly (Ed., 1996), R. Eisler

(1995), T.G. Frantz (1993), A. Gutman and D. Thompson (1996), R.B. Reich (1991), M.J. Sandel (1996), P. Slater (1991), Yonassan Gershom (1992). A research team on social design in the International Systems Institute completed several papers to appear - B.H. Banathy (1997, 1989a), G. Alexander (1997), Matthew A. Shapiro (1997). One special optimisation problem is taken out of this vast field and, with its experience and wisdom, attempts are made to create solutions and match them to global, social and information conditions.

2. A SYSTEMIC VIEW ON THE PROBLEMS

Systems theory of chaos is misunderstood or wrongly applied in this connection. Each individual's intention is personal and hence stronger and more fruitful than a number of intentions with different directions summed up in a total of zero and added to an enforced vision.

It is fact that in times of crises people, good in their souls, are sacrificing a lot to make their children's, their ones', their friends' lives more bearable or just to survive. Individually earned small amounts of money can already be invested "for a good deed", and their owners would like to do so, and they already know how to do so.

A very important barrier is misundertsood (or neglected) by the authorities. The usual beginning is to start an one person activity, to hire labour for one single task (amending, repairing work, etc.), thereafter to make an oral contract for several hours with some periodicity for a greater project. An official appointment cannot be made because there is not enough money to spend, not enough stability to keep one's word, often not enough amount of needed work to bother about legalising a working place. Taxes and insurance are not paid - both contractors are dissatisfied in these cases nowadays in Bulgaria. This means that the legislative soil is already prepared. But not yet fully used. Consequently, the effect of bettering life standard is smaller, discontent emerges on a wide basis, many smaller herds of conflict arise, widen, etc.

In the European Union this kind of work is tolerated as "black labour market". Why ?

The essence lies in the high numbers of frequency and reiteration of the phenomenon. Moreover, it may be a small niche, but with a great potential in its vacuum - a germ put there would grow acceleratedly. There are live examples for it in another context.

2.1 *Can a Job Unit Be Smaller or Greater Than One?*

Now, there is no place for a phenomenon giving small but numerous effects with a huge sum in the labour-investment system. The system's premise is that the smallest "unit of labour" is a full occupation for at least one year. The law does not limit to it. Community's Labour Offices try their best. But their vision embraces only unpleasant and occasional hourly work for which one has to queue early in the morning every day again and again - no reasonable optimising of behaviour is possible - the vision is for workers with almost no education, no training. Rational thinking and optimising behaviour within the limits of the existing system comes to this result both in the role of worker and employer-investor.

It is fact that the first steps of creating an enterprise start by turning an "occasional help", a "small unit of labour" (SUL) to a "full time employee", accumulating further "occasional help" units. And this is the soundest way to develop. And it is not always true, that some steps can be overridden by investing a bigger amount of money just at the beginning. But it is true that sometimes investing greater amount of money at the very beginning incorporates unhealthy elements in the system. Because great systems are not organically adapted to its function and environment but taken as analogon from previous models. I could enumerate a long chain of thoughts from this point on, but let us turn our attention back to the positive side of life.

What needs a civil society in order to manage this phenomenon, what lacks the existing system of thinking and practice, and legislation, and information infrastructure ? What interactions and securities were neglected ? What will be the parameters for optimising it, (some of) that failed up to now ? What (parts of) roles has to be added to the usual infrastructure of a civil society ?

Theories of transition offer a military or police dictatorship, which absorbs these free germs of wealth and good intention, activity, "wish to invest work" by transferring it to unpaid enforced work, which, of course, is badly done, if at all. The military type of system is well functioning, if it is well defined, but it is its essence not to allow any chaos, even a pregnant one. Moreover, it fears any intervention in a well defined system. And fear is the worst advisor of all. Another wrong way to absorb germs of wealth is the inflation, thus the mechanism of generating crime is still accelerated.

2.2 *SUL and Other Global Problems*

This is not a trivial problem because a SUL cannot keep a person alive, she/he has to have another source of income and survival support. A very important principle is that every person at any moment should feel free to either help developing a SUL to a bigger one or to look for another SUL, or to combine several SULs, to order them in time and space according to her/his training, learnt knowledge, family duties and burdens, etc., or to a full time job. Especially, if for some reasons, the current one does not satisfy her/him. Such a system may help to close loosing enterprises in a most natural way. It happens occasionally.

The "black labour market" functions quite well in big cities, because of its simple optimisation routine, there are minimal transport costs, there are good probabilities to reach necessary information through common human relations outside any legal public information systems. People are just walking from one opportunity to another, meeting in cafes or other "special" places - the effort to find the needed is bearable and could stay unnoticed or be tolerated by the current employer.

The problem of this kind of optimisation turns out to coincide in its effects with one of the global problems - the overgrowth of cities. And, maybe, in its causes, at least partly. And, maybe, in its partly solution. A systemic view over "black labour market", SUL, "good will", wrong investment, analogy of models, information infrastructure, etc. shows the need "to put the ends together", to create a mechanism of informing and optimising on an individual level conform with the possibilities of the Global Information Society, against negative trends of development in other global problems. There is a lot of scientific sources on the global problems, e.g. the problem of mobility vs. distributed deployment. Some of them were worked through on an other occasion by the author.

Mathematically seen this kind of optimisation leads to a heavy problem to solve. One of the ways to gain time (it is still done by means of trial and error mainly) and money (for its solution, which is mastered by enormous amounts of expenditures), is to make the interactive information system as fast and as flexible as possible, and, before all, to follow the natural tracks of interconnections in the existing "black", illegal system.

From the social view point it is a problem of social design. Some useful bibliography is sampled below.

2.3 *Who Are the Agents in the System ?*

People there are willing to pay for small amounts of work on hourly basis - this situation is specific for a society, which is gradually, but slowly coming up from a severe crisis. Seasonal work has very old tradition in Bulgaria. Men (but also women nowadays) go to the cities or to intensive agricultural regions in the summer looking for short term jobs. That is the reason why I take it as an example for an "occasional help" unit of labour - a SUL.

Thus the seasonal work has about ten parties building a system: 1) the worker, 2) the private person offering jobs on hourly basis - the employer, 3) the population in the region (including officials in neighbouring firms and institutions, fearing strangers for good reasons, 4) enterpreneurs, 5) people engaged in trade and services in the region, including those on seasonal basis, 6) community officials, e.g. for state taxes, 7) people offering seasonal accommodation, 8) journalists in mass media, 9) local police officers, 10) Labour Office employees. This office is the only jobber on the labour market in Bulgaria, recently some serious private firms are emerging.

2.4 *Conditions in Bulgaria*

The Labour Offices are computerised (to a certain extent), as well as the private firms. They have data bases, they communicate throughout the country among each other by phone. In a document describing a project proposal are enumerated measures enhancing: 1) jobbing before the beginning of the season and enabling the worker's family to choose the region(s) sequence ahead, 2) reducing the out-of-job hours, days at the temporary job location to a minimum, 3) regulating the employment according to the natural interests of all 10 agents in the system, 4) explanation and education by the mass media.

A great number of people in Bulgaria would still have a permanent job (seasonally independent), interrupt it during the season for a better payment, and then return back home to the permanent one. Here one must have in mind, that permanent job salaries in Bulgaria hardly cover the minimum existence of families (see negative birth trends; there are even more crucial statistics showing the gradual extermination of the nation.).

An applied research project could analyse the social conditions under which occupation optimising can be helpful both to develop society, to stop deterioration of educated people and to speed up economic development. There are several reasons to start investigations in Bulgaria:

1. Bulgaria has an ancient culture of ruling, governing, living according to (changing) natural facts, premises, conditions.

2. Bulgaria has ancient traditions in communicating without means and tools, that is on the human level and - in this way - in communication systems thinking.

3. Bulgaria is a mountainous country and is feeling that wired and wireless communication infrastructure has to replace many of the channels, which countries with an easier geography do by personal transport.

4. Bulgaria's people posses a unique combination of agricultural and intellectual traditions and wisdom, as well as modern high tech knowledge.

5. Bulgaria has ancient tradition in seasonal labour mobility.

6. Bulgaria is a relatively poor country, due to governed, imposed from the outside economic mischief; it has debts, and very few (for now) possibilities to produce goods for the international market.

7. Bulgaria has a well trained class of intellectuals in particular for information and communication technologies.

8. Bulgaria is looking for a new economic structure according to all its historical and geographical advantages and willing to work it out on all levels of the society on a high level of conscience.

9. Last, but not least, Bulgaria's population made its country a core of regional stability and is willing to apply its wisdom to make it a wealthy one.

Certain control and security measures are necessary for the whole system and have to be designed simultaneously with its design.

3. CONCLUSION

In all, the investigation should aim at designing a regulated highly flexible labour market respecting interests and people. Is there a solution neglected up to now ? Is there a chance for people intellectually well off ? Is their only perspective to deteriorate to "occasional labour force"? Can intellectual work be more than a hobby without being a full time job ? Is there, at the sunrise of the era of Global Information Society, a technique and methodology to surmount several global problems at a time ?

REFERENCES

Alexander, G. (1997). Community: The corner stone of building a public philosophy of cultural democracy an democratic culture. *Asilomar conversations.*

Banathy, B.H. (1989a). Design: A journey to create the future: A map of the journey. *1st Pacific Rim Fuschl Conversation.*

Banathy, B.H. (1989b). The design of evolutionary guidance systems. *Systems Research*, **6(4)**, 289-295.

Banathy, B.H. (1996). *Designing social systems in a changing world: A journey toward a creating society.* Plenum Press, New York.

Banathy, B.H. (1997). Knowledge base for the design of an ideal society. *Asilomar conversation.*

Bellah, R.N., Madsen, R., Sullivan, W.M., Swidler, A., Tipton, S. (1991). *The good society.* Alfred A. Knopf, New York.

Dahl, R. A. (1989). *Democracy and its critics.* Yale University Press, New Haven.

Eberly, D.E. (Ed.) (1996). *Building a community of citizens.* University Press of America, Lanham.

Eisler, R. (1995). *Sacred pleasure.* Harper Collins, San Francisco CA.

Follett, Mary.P. (1918). *The new state.* Longman and Green Publishing, MA.

Frank Michelman. (1988) Law's Republic. *Yale Law Review* **97**, 1493-1537

Frantz, T.G. (1993). Jumping out of existing system during design genesis: Penetrating the anxiety barrier. In: *Comprehensive systems design: A new educational technology* (C.M. Reigluth, et al Eds.),., Vol. 95, Springer-Verlag, NATO ASI Series F: Computer and Systems Sciences, New York.

Gutman, A., Thompson, D. (1996). *Democracy and disagreement.* Belknap Press, Cambridge, MA.

Jan Smuts (1926). *Holism and Evolution.* 345 p.

Martin Buber (1958). *Paths in Utopia.*

Reich, R.B. (1991). *The work of nations.* Simon & Schuster, New York.

Ronald Arnett (1986). *Communication and Community: Implications of Martin Buber's Dialogue,* Southern Illinois University Press.

Sandel, M.J. (1996). *Democracy's discontent.* Belknap, Harvard University Press, Cambridge, MA.

Shapiro, Matthew A. (1997). The paradox of the procedural and the promise of the designing. *Asilomar conversations.*

Slater, P. (1991). *A dream deferred.* Beacon Press, Boston.

Yonassan Gershom (1992). *Beyond the Ashes.* ARE Press, Va. Beach, VA

GAINS FROM INTERNATIONAL MACROECONOMIC POLICY COORDINATION UNDER ALTERNATIVE OBJECTIVE FUNCTIONS

Gottfried Haber, Warwick J. McKibbin, Reinhard Neck

Department of Economics, University of Klagenfurt, Austria
Australian National University, Australia & The Brookings Institution, Washington, DC
Department of Economics, University of Klagenfurt, Austria

Abstract: This paper investigates whether coordinated macroeconomic policy reactions to a negative productivity shock give better results than non-coordinated policies and passive adaptations to a shock. The framework used is the McKibbin-Sachs Global Model of the world economy, which incorporates rational expectations. The USA, Japan and Germany are regarded as players in a dynamic game, using money supplies and government consumption as strategic variables. The implications of different specifications of the objective functions are investigated. It is shown that in general, cooperative policies give results that are superior to non-cooperative Nash equilibrium solutions. The gains from active policy-making and from cooperation depend strongly on the assumptions made about the objective functions, in particular about the weights given to different target variables. *Copyright © 1998 IFAC*

Keywords: economics, modeling, dynamic models, simulation, game theory, cooperation, coordination, equilibrium, quadratic performance indices, sensitivity analysis.

1. INTRODUCTION

One of the central topics in the recent literature about international policy cooperation is the question of whether coordinated stabilization policies yield better results with respect to macroeconomic target variables than uncoordinated discretionary policies or fixed policy rules. Most theoretical and empirical studies have found that cooperative policies are superior to non-cooperative ones, but the gains from international macroeconomic policy coordination may be rather small. However, in models with rational expectations it is possible that cooperative policies result in outcomes which are worse than those of non-cooperative policies. The same may be true when there are more than two policy-makers with different objective functions and possibilities for strategic interactions.

In this paper, coordinated macroeconomic policy reactions to a global negative productivity shock are compared to non-coordinated policies and passive adaptations to such a shock. The framework used is the McKibbin-Sachs Global Model of the world economy, which incorporates rational expectations of private-sector agents. The USA, Japan and Germany are regarded as players in a dynamic game. Policy-makers in these three countries are assumed to have different objective functions and use monetary and fiscal policies as strategic instruments. In particular, it is examined how results of non-cooperative and cooperative policies depend on particular assumptions about the policy-makers' objective functions. Therefore, non-cooperative and cooperative solutions of the dynamic policy game under alternative assumptions about the discount factor (the rate of time preference), the planning horizon and the weights in the objective functions given to different target variables, such as inflation, employment, growth, the current account and the budget deficit, are determined. The results show that in most cases, cooperative policies give results that are superior to non-cooperative equilibrium solutions. However, the gains from cooperation and in particular their distribution depend strongly on the weights which policy-makers attach to different target variables. For policy purposes, this implies that a careful specification of the objective function is essential in an analysis of possible gains from international policy cooperation.

2. GAME-THEORETIC ANALYSES OF INTERNATIONAL POLICY COORDINATION

Early applications of game theory to international economics (Cooper, 1969; Hamada, 1976) used static models to show the inefficiency of non-cooperative equilibrium strategies for policy-making in a two-country framework. This can be interpreted to mean that cooperation between policy-makers of different countries is advantageous for at least one and often both countries; however, some coordination is required because efficient cooperative solutions are usually no equilibria. During the eighties, this research has been extended to dynamic economic models, using either repeated or dynamic game theory. In addition, empirical evidence for the theoretical predictions about the effects of policy coordination has been looked for. For surveys on this literature, see Currie *et al.* (1989) and McKibbin and Sachs (1991, ch. 7); see also Currie and Levine (1993) for a more theoretical treatment.

Most theoretical applications of game theory so far have been confined to two-country models. In this context, the advantages of policy coordination can be shown under fairly general conditions. Simulations of theoretical models with numerical values assigned to the parameters as well as those of macroeconometric models have also obtained quantitative estimates of the gains from international policy coordination. In general, these turn out to be rather small, although their magnitude depends on the amount of spillovers present in the different models, on the nature of the shocks considered, and on the objective functions assumed for the policy-makers (see McKibbin, 1997). Moreover, the distribution of the gains from coordination among the countries involved has been investigated. Usually an asymmetric distribution emerges, which can be one reason for the political difficulties in bringing about and sustaining cooperative agreements.

In contrast to the simpler game-theoretical models of policy coordination, some more elaborate studies have shown that cooperation may be disadvantageous. This may happen when more than two policy-makers are taken into account. In this case, the possibility of coalitions arises, and a much greater variety of results can emerge. A similar ambiguity arises when problems of time-inconsistency and lack of credibility of policies are present. This is often the case when private agents are assumed to have rational or forward-looking expectations. In particular, Rogoff (1985) has shown for a rational-expectations model of open economies that international policy coordination can reduce welfare if coordination makes it easier for rational policy-makers to engage in inflationary monetary expansion. Such a situation can be interpreted as a coalition of different countries' policy-makers against rational, strategically acting private agents in the two countries. A large literature has originated from Rogoff's result, mostly concentrating on the requirement for policy-makers to build up reputational mechanisms to prevent counter-productive policy coordination. Some authors have argued for "fixed" rules or other "simple" policy designs, but the theoretical foundations of these proposals are not generally accepted. The empirical evidence presented in the large study by Bryant *et al.* (1994) suggests that some simple rules perform well in empirical models.

Despite the large amount of research undertaken to date, the question of whether there are gains from international macroeconomic policy coordination and, if so, whether they are quantitatively important remains open in the presence of more than two countries and of private agents with rational expectations. The same holds for the distribution of these gains among the countries. Therefore, there is some need for additional game-theoretic analysis of these issues, preferably based on empirical global macroeconomic models. In this paper, some results from such a study are presented, using the McKibbin-Sachs Global Model and a dynamic game framework with three active policy-makers and with private agents having forward-looking expectations. In particular, the dependence of the results of non-cooperative and cooperative macroeconomic policies on the specification of the policy-makers' objective functions are investigated. This extends and modifies results obtained in McKibbin *et al.* (1996).

3. THE McKIBBIN-SACHS GLOBAL MODEL

The McKibbin-Sachs Global Model (MSG2 Model) is a dynamic, intertemporal, general-equilibrium model of a multiregion world economy. It is based on microeconomic foundations by assuming that economic agents maximize intertemporal objective functions. The model exhibits a mixture of classical and Keynesian properties: expectations are assumed to be formed in a rational way, but various rigidities are taken into account by allowing for deviations from fully optimizing behavior. In particular, nominal wages are assumed to adjust slowly in the major industrial economies (except for Japan); due to this wage stickiness, extended periods of unemployment can be present in these economies. Nevertheless, the model solves for a full intertemporal equilibrium in which agents have rational expectations of future variables. As a model with theoretically constrained long-run properties, it can display how the short-run adjustment of the world economy to exogenous shocks depends upon the long-run adjustment.

The theoretical structure of the model as well as a listing of its equations is given in McKibbin and Sachs (1991) and additional documentation can be found on the Internet at WWW.MSGPL.COM.AU; here only some of its theoretical features are pointed out which make it particularly well suited for analyzing adjustments to exogenous shocks. First, the long run of the world economy is well determined, being driven by a neoclassical growth model, with exogenous technical progress and population growth. In the short run, on the other hand, the dynamics of the global economy towards this growth path is determined both by Keynesian rigidities in the goods and labor markets and by optimal decisions, conditional on expected future paths of the world economy. Thus, the model takes into account both theoretical considerations of long-run effects of shocks and short-run dynamics towards these long-run outcomes based on historical experience, with expectations formation providing a link between the long-run outcome and the short-run adjustment.

Secondly, the MSG2 Model is a fully specified dynamic general-equilibrium model incorporating both the demand and the supply sides of the major industrial economies. Stock-flow relations are carefully observed, and intertemporal budget constraints are imposed. Intertemporal budget constraints and forward-looking expectations require that all outstanding stocks of assets must be ultimately serviced. Underlying growth of Harrod-neutral productivity plus growth in the labor force is assumed to be 2.5 percent for each region. Given the long-run properties of the model, the world economy settles down to the 2.5 percent steady-state growth path after any set of initial disturbances.

Thirdly, asset markets are efficient as asset prices are determined by intertemporal arbitrage conditions and rational expectations. Asset prices are tied down by the imposition of intertemporal budget constraints. The long-run behavior of the model depends on stock equilibrium rather than flow equilibrium. Asset prices stabilize in real terms, once desired ratios of asset stocks to GDP are reached. The short run of the model behaves in a similar way as the basic Mundell-Fleming model under flexible exchange rates and high capital mobility; however, the future paths of the world economy are important in the short run because of the forward-looking behavior in asset and goods markets. The assumptions of rational expectations in financial markets and of partially forward-looking behavior in real spending decisions allow for incorporation of the effects of anticipated policy changes. As a consequence, every simulation requires that the entire future sequence of anticipated policies must be specified.

Finally, the supply side of the model is specified in an internally consistent manner. Factor input decisions are based in part on intertemporal profit maximization by firms. Labor and intermediate inputs are determined to maximize short-run profits, given a stock of capital that is fixed within each period and adjusted according to a Tobin's q-model of investment, where Tobin's q evolves according to a rational-expectations forecast of future after-tax profitability. The wage-price dynamics, on the other hand, is specified on the basis of empirical evidence concerning differences in the wage-price processes in the United States and Europe on the one hand and Japan on the other, resulting in different degrees of wage and price stickiness in these regions.

The version of the MSG2 Model used in this paper, called MSGR44A, consists of models of the following countries and regions: United States, Japan, Germany, United Kingdom, France, Italy, Austria, the rest of the European Monetary System (REMS), the rest of the OECD (ROECD), Central and Eastern European economies (CEE), non-oil developing countries, oil-exporting countries, and the former Soviet Union. For the latter three regions, only the foreign trade and external financial aspects are modeled, whereas the industrial countries and regions are fully modeled with an internal macroeconomic structure. Although the basic theoretical structure for all industrial regions is the same, institutional differences are taken into account, especially in modeling labor markets. The exchange rate arrangements of the European Monetary System are modeled by assuming the EMS to be a DM-zone.

In contrast to macroeconometric world models, the MSG2 Model is fitted to macroeconomic data by a mix of calibration techniques for computable general-equilibrium models and econometric time-series estimates. Behavioral parameters taken from econometric studies and data (for 1992) for macro aggregates were combined with steady-state relations in the model to generate other data. The year 1992, for which actual data were replicated, is not regarded as representing a steady state of the model but a point on the stable adjustment path towards the steady state, hence not all steady-state relations are assumed to hold for that year. The model is solved in linearized form, with the linearization taking place at a point in time (1992, in our case) instead of along some reference path. The baseline is updated to 1997.

4. POLICY REACTIONS TO A PRODUCTIVITY SHOCK - SIMULATION LAYOUT

In this section, the simulation layout and the response of the global economy to a negative productivity shock are described. In addition, several simulations with the MSG2 Model have been run to explore the sensitivity of the results concerning the different scenarios. These results will be presented in the next section.

In the simulation experiments, a negative global productivity shock is imposed under different assumptions regarding economic policy. There are three basic types of simulations: "no-policy", "co-operative" and "non-cooperative".

First, a baseline solution of the dynamic model has to be calculated. This baseline solution can be seen as a stable adjustment path towards the long-run growth path of the model. Therefore, there are good reasons to interpret this baseline solution as an optimal path of the economy. When calculating this baseline of the model, the exogenous variables (in the broadest sense, including the instrument variables) are kept at constant values or constant growth rates. This projection serves as a benchmark for the economic performance of each policy-maker and for the world economy as a whole.

The next step is to simulate different shocks on the exogenous variables and to analyze the time paths of selected key variables. To compare the welfare effects of different policy actions, a single measure of economic performance is needed. Such a measure can be the intertemporal welfare losses due to the simulated shock. To calculate these welfare losses, an objective function has to be specified. For computational ease, an additively separable quadratic welfare loss function has been chosen. The welfare losses Ω_t in each period t are equal to the sums of the weighted (λ_i) quadratic differences between the actual values τ_i and the optimal values τ_i^* for each of the i target variables:

$$\Omega = \sum_{t=1}^{T}(1+r)^{-t}\Omega_t\,,\ \Omega_t = \sum_i \lambda_i\left(\tau_i - \tau_i^*\right)^2 \qquad (1)$$

In order to take into account the dynamic structure of the overall welfare losses, the welfare losses in each period have to be discounted to their present values (using the rate of time preference r, which is assumed to be 10 percent in the base simulations) and to be summed up over the time horizon T (103 years in the simulations, from 1998 to 2100) to get the total welfare loss.

The target variables in the following simulations are inflation, employment, real GDP, the current account and the budget deficit. For our purpose, in the base simulations all target weights are set equal to 0.2, producing an equally weighted objective function which is standardized, as the weights sum up to 1. As mentioned above, the baseline values of the target variables are considered as their optimal values. Note that this implies that the welfare losses in the baseline scenario have to be zero, which is another reason for using this baseline as benchmark for all simulations.

For the scenarios without active macroeconomic policy, it is assumed that the instrument variables of the policy-makers in all countries are set at the same values as in the baseline solution ("no-policy" or fixed-policy simulations). In this case, the calculation of the welfare effects is straightforward: First, the dynamic model is solved subject to the exogenous shock. Then, the values of the objective functions are calculated.

In the simulations with dynamic optimization, the policy-makers of the USA, Japan and Germany are considered as players in a dynamic game. The players set the values of their respective instrument variables in each period. In the "non-cooperative" case, they do so by minimizing their welfare loss functions subject to the dynamic model and given the optimizing behavior of the other players. This leads to a Nash-Cournot equilibrium of the dynamic game. In the "cooperative" case, a joint welfare loss function, which is a weighted sum of the individual objective functions, is minimized subject to the dynamic model. This is equivalent to assuming a global dictator, who minimizes overall welfare losses of the players involved, and can be interpreted as the result of an agreement between the policy-makers of the three countries involved (the "G3"). It corresponds to the collusive solution in game theory, because all players get equal weights in the joint objective function for the base simulations.

The MSG2 Model assumes rational expectations for private-sector agents; hence, some complications arise for the resulting dynamic games. Either the entire future paths of all instrument variables (open-loop policy) or a policy rule for the instrument variables could be calculated (closed-loop policy) as a solution for the dynamic game. Next, the problem of time-inconsistency has to be taken into account. Time-inconsistency means that at a future time point, re-optimization results in time paths of the instruments which are different from the optimal open-loop policy. The presence of forward-looking private agents can be interpreted as the presence of another (implicit) player in the dynamic game. Therefore, the solution of a standard optimum control problem may not be carried out.

The solution algorithm DYNGAME, which is used to solve the MSG2 Model, calculates strongly time-consistent, closed-loop policy rules; hence its solutions do not suffer from the time-inconsistency problem. This has to be kept in mind when interpreting the results of the dynamic simulations involving strategic policy optimization.

The productivity shock can be interpreted as a temporary inward shift of the production possibility frontiers of all countries. It may be caused, for example, by an environmental catastrophe resulting in a reduction of the supply of intermediate goods required for producing industrial goods, or by another exogenous reduction in total factor productivity. In particular, total factor productivity is assumed to fall by 4 percent in 1998, 3 percent in 1999, 2 percent in 2000, and 1 percent in 2001 as compared to the baseline of the model.

The main focus of our analysis will be on the USA, Japan and Germany, which will be denoted as the "G3". In the fixed-policy scenario, real GDP and its components are significantly below their respective values in the baseline run. At the same time, the price levels for the G3 are higher. Thus, this scenario results in a stagflationary process. Policy-makers in the G3 are confronted with the typical trade-off between reducing inflation at the cost of cutting GDP and improving GDP at the expense of even higher inflation.

In the "non-cooperative" scenario, the USA adopts contractionary monetary policies in 1998 (period 1) and contractionary fiscal policies in 1998 and 1999. This reduces inflation. Afterwards, both economic policy instruments are above their baseline values, stimulating GDP. In the first periods, GDP and its components are lower than in the "fixed-policy" simulation, but they are higher afterwards, showing more rapid convergence to the long-run path of the economy. Japan implements restrictive monetary policies in 1998 and 1999 and expansionary fiscal policies in all periods, while Germany adopts restrictive monetary policies in 1998, 1999 and 2000 and restrictive fiscal policies in 2002 and afterwards.

On the other hand, the "cooperative" scenario prescribes other strategies for the policy instruments. The USA implements expansionary monetary policies from the beginning and react in a less restrictive way with respect to fiscal policy in the first periods, but also in a less expansionary way afterwards. Japan applies less expansionary fiscal policies and less restrictive monetary policies in the first periods. Germany now uses expansionary fiscal policies in all periods and tends to adopt less restrictive monetary policies, too. The differences in the target variables have a complex structure and reverse their signs frequently. Due to lack of space, these effects are not described here in detail.

Table 1 Values of the standardized objective functions, productivity shock to all model regions

Region	"No-policy" scenario	"Non-cooperative policy"	"Co-operative policy"
USA	19.61	14.80	15.59
Japan	11.00	15.27	11.41
Germany	21.87	25.83	19.55
Sum	52.48	55.90	46.55

The results in Table 1 show that the "cooperative" scenario is clearly superior to the "non-cooperative" scenario in terms of welfare losses for the world as a whole but not for the USA in particular. Note that the dynamic feedback Nash equilibrium (the "non-cooperative scenario") is inferior to the fixed-policy scenario.

5. SENSITIVITIES WITH RESPECT TO TIME PERFERENCE, TIME HORIZON AND WEIGHTS

The simulation results described in the above section do not depend on the specification of the model only, but are also subject to the specification of the objective functions of the dynamic players. Several simulations have been run with systematic variations of the weights attached to the target variables, the weights of the individual countries, the discount rates (rates of time preference) and the time horizon of the simulations.

In order to evaluate the sensitivity with respect to the discount rates, three alternative simulation sets have been considered. Starting with the base simulation set from the previous section, alternative rates of time preference for the policy-makers are evaluated, namely discount rates of 0 percent, 10 percent (the value for all other experiments) and 10,000 percent. In the first case, all periods are of equal importance to the decision-makers. This makes only sense for a temporary shock and a model that returns to a long-run equilibrium after the initial shock. As this is the case for the MSG2 Model, these zero weights were selected as one extreme case for the sensitivity analysis. On the other hand, 10,000 percent is an approximation for an infinite discount rate and implies that only the first period matters.

The results of these simulations show that even unreasonably high variations of the discount rates do not lead to substantial differences either in the instruments or in the targets. The "non-cooperative" scenarios exhibit less variation than the "cooperative" scenarios. For the "non-cooperative" and 10,000 percent case, monetary policy is more restrictive or less expansionary. That is true for the USA, Japan and Germany. When there is no discounting at all, the USA act in a more expansionary or less restrictive way in the first periods, while the opposite is the case for Japan. In the "cooperative" case, slightly different results emerge.

Next, the sensitivity of the results with respect to the values of the welfare losses is evaluated. Tables 2 and 3 display the values of the objective functions (with the 10 percent discount rates used for the calculations) for the three simulations with different discount rates. Note that the 10 percent discount rate is assumed to be the "true" discount rate. Even in the case of severe misspecification (the other two discount rates), the variation coefficients (VCs) are below 0.12 for the "non-cooperative" scenarios. The VCs for the "cooperative" scenario are higher, but still below 0.22. Hence the sensitivity of the results with respect to the discount rates is rather small, though the cooperative layout tends to react more sensitively.

Table 2 Values of the standardized objective functions, different discount rates, non-cooperative

Discount rate	0%	10%	10,000%
USA	14.72	14.80	15.65
Japan	15.21	15.27	15.96
Germany	25.52	25.83	28.35
Sum	55.45	55.90	59.96

Table 3 Values of the standardized objective functions, different discount rates, cooperative

Discount rate	0%	10%	10,000%
USA	16.79	15.59	13.94
Japan	11.29	11.41	12.09
Germany	21.88	19.55	17.56
Sum	49.96	46.55	43.59

As the time horizon for the simulations can be seen as an approximation of the infinite time horizon, the results of the dynamic optimizations are extremely robust against variations of the number of periods in the simulations as long as convergence is given. As a rule of thumb, there should be at least 30 periods for the MSG2 Model to arrive at a stable solution. The time paths of all model variables are nearly unchanged in comparison with the base scenarios from the previous section.

In these experiments, the time horizon is set to 30, 103 and 250 periods. The standard deviations of the values of the objective function are below 0.004 for the "non-cooperative" and the "cooperative" scenarios; variation coefficients are below 0.003. There are definitely no differences in the time paths for all variables for the 103 and 250 periods experiments. Even in the "short term" experiments, only very small deviations appear.

When performing optimal policy calculations, one of the central questions is how to set the weights in the objective functions. There are two different types of weights: the weights for each individual target variable and the weights for each player in the "cooperative" scenarios. It is not straightforward to compare different weighting schemes in simulations, because changing the weights in the objective functions leads to completely new specifications of the objective functions. Of course, this problem applies to all experiments that are based upon variations of the functional form or the parameters of the welfare loss function. Therefore, a "true" objective function is assumed for the sensitivity analysis. When considering different weighting schemes for the targets, it is interesting whether welfare effects are similar in the simulations.

First, the effects of systematic variations of the target weights are analyzed. Table 4 shows the different weighting schemes. The same weights are assumed for the objective functions of the three countries in each case.

Table 4 Weighting schemes for the target weights sensitivity analysis

	Inflation	Employ-ment	GDP	Current account	Budget deficit
a)	0.20	0.20	0.20	0.20	0.20
b)	0.40	0.15	0.15	0.15	0.15
c)	0.15	0.40	0.15	0.15	0.15
d)	0.15	0.15	0.40	0.15	0.15
e)	0.15	0.15	0.15	0.40	0.15
f)	0.15	0.15	0.15	0.15	0.40

The first case (a) corresponds to the base simulation from the previous section. As mentioned above, in these experiments the variations of the target values have to be interpreted carefully. In spite of that, the effects on the targets and instruments are unambiguous: The calculations show superior results for those target variables which get the higher target weights as compared to the equal-weights simulation.

Table 5 Values of the standardized objective functions, different weighting schemes, non-cooperative

Weighting scheme	a	b	d	e	f
USA	14.80	19.58	18.07	17.19	10.18
Japan	15.27	16.19	19.31	33.86	9.49
Germany	25.83	35.12	27.72	22.40	16.94
Sum	55.90	70.89	65.10	73.45	36.61

Table 6 Values of the standardized objective functions, different weighting schemes, cooperative

Weighting scheme	a	b	d	e	f
USA	15.59	17.73	18.43	23.99	10.63
Japan	11.41	10.78	16.61	18.67	8.40
Germany	19.55	20.18	23.08	10.28	15.61
Sum	46.55	48.69	58.12	52.94	34.64

Tables 5 and 6 show the resulting welfare losses under the objective functions with weights taken from Table 4 for the "non-cooperative" and the "cooperative" scenarios, respectively. The results for the weighting scheme c) are not reported as the model did not converge to a reasonable solution under this scheme in the cooperative case. The reason for this is the neoclassical feature of the MSG2 Model which provides no easy possibility to influence employment by demand-side policies. If this target variable gets a high weight, erratic behavior of instrument variables and instability of the solution may arise. In all other cases, the "cooperative" scenario gives lower overall welfare losses than the "non-cooperative" one, although the former need not dominate the latter (i.e., one country may be better off in the "non-cooperative" scenario). It is interesting to note that only under the weighting scheme b), the cooperative solution is dominant; under the other weighting schemes of Table 4, the USA is better off in the non-cooperative solution and hence has no incentive to enter a cooperative agreement. If policy-makers strongly dislike inflation, cooperation will become more likely in this model, *ceteris paribus*.

It must be concluded that the values of the instruments, the values of the targets, the structure and the distribution of the welfare gains strongly depend on the weights attached to the targets. In several simulations, there appear even sets of target weights for which no stable policy rules exist. Therefore, the existence and stability of time-consistent policy rules also depend on these weights. Especially high weights for unemployment are likely to produce unstable results.

Next, different weights for the players in the "cooperative" simulations were considered. Lack of space precludes a detailed discussion of the results of these simulations. The main result is that the values of the instruments and the targets strongly depend on the weights of the dynamic players. If, for example, a "true" objective function with equal weights is assumed, deviations from the "truly" optimal values resulting from optimizing the misspecified common objective function lead to substantial welfare losses. This problem becomes the more severe the higher the weight for a single country is, i.e. the sharper the discrepancy between the "true" and the misspecified weights is.

Apart from the strong impact of misspecification of the objective function, the simulations show that very asymmetric weighting schemes lead to high welfare losses in the countries with the lower weights, even in terms of their weighted welfare losses. It can be concluded that the results of the dynamic optimization problem strongly depend on the cooperative weights, too.

6. CONCLUDING REMARKS

The simulations presented in the above sections show that cooperative policies in general give better results than non-cooperative policies. The sensitivity of the results with respect to the chosen time horizon and the discount rates in the objective functions is rather low. However, the weights for both the target variables and the players (in the cooperative case) have a strong impact on the design of monetary and fiscal policies and thus lead to completely different results. For this reason, much care has to be exercised when specifying these weights. Misspecifications of policy-makers' preferences with respect to different targets may lead to substantial welfare losses for all players.

ACKNOWLEDGEMENT

Financial support from the Jubiläumsfonds der Oesterreichischen Nationalbank (project no. 6917) and from the Ludwig Boltzmann Institute for Economic Analysis, Vienna, is gratefully acknowledged.

REFERENCES

Bryant R., P. Hooper, and C. Mann (1993). *Evaluating Policy Regimes: New Research in Empirical Macroeconomics*. Brookings Institution, Washington, DC.

Cooper, R.N. (1969). Macroeconomic policy adjustment in interdependent economies. *Quarterly Journal of Economics*, **83**, 1 - 24.

Currie, D.A., G. Holtham and A. Hughes Hallett (1989). The theory and practice of international policy coordination: does coordination pay? In: *Macroeconomic Policies in an Interdependent World* (R.C. Bryant *et al.*, Eds.), pp. 14 - 46. IMF, Washington, DC.

Currie, D. and P. Levine (1993). *Rules, Reputation and Macroeconomic Policy Coordination*, chapter 2. Cambridge University Press, Cambridge.

Hamada, K. (1976). A strategic analysis of monetary interdependence. *Journal of Political Economy*, **84**, 677 - 700.

McKibbin, W.J. (1995). *DYNGAME: An Algorithm for Solving Dynamic Games in Rational Expectations Models*. McKibbin Software Group Inc, Arlington, VA. (WWW.MSGPL.COM.AU)

McKibbin, W.J. (1997). Empirical evidence on international economic policy coordination. In: *Handbook of Comparative Economic Policies*, Vol. 5: *Macroeconomic Policy in Open Economies* (M. Fratianni, D. Salvatore and J. von Hagen, Eds.). Chap. 5, pp. 148-176. Greenwood Press, Westport, Conn.

McKibbin, W.J., R. Neck and G. Haber (1996). On the gains from international macroeconomic policy coordination. In: *Supplementary Ways for Improving International Stability* (P. Kopacek, Ed.), pp. 53-59. Pergamon, Oxford.

McKibbin, W.J. and J.D. Sachs (1991). *Global Linkages*. Brookings Institution, Washington, DC.

Rogoff, K. (1985). Can international monetary policy cooperation be counterproductive? *Journal of International Economics*, **18**, 199 - 217.

SWAY OF NEW TIME: GENERAL LINES

Frederick O.Kile*, Peter P.Groumpos, Florian G.Filip***, Svetlana Novikava******

**) Microtrend, USA, e-mail: 102610.2345@compuserve.com*
***) Laboratory for Automation&Robotics, University of Patras, Greece,*
e-mail: groumpos@ee.upatras.gr
****) Research Institute for Informatics, Romania,*
e-mail: filipf@roearn.ici.ro
*****) Hierarchical Multilevel Systems Laboratory,*
Institute of Mathematics&Cybernetics, Belarus,
e-mail: imc@org.by

Abstract: The paper considers the main characteristics of sway (design&control&learning) process with new cybernetical means - hierarchical networks of States and unions of States. The hierarchical networks are defined in agreement with real world construction which connects the strata (levels, times) created by its history and new ones whose hazy images belong now to the area of thoughts. The new mechanism of mathematics (aed theory) allows to turn these images into symbols (figures) convenient in sway process. The strategy of new time constructing acquires in suggested symbols of aed theory the lines of mathematical records which have identical understanding in wide scientific and practical ranges able to change this strategy. Owing to that the high hazard of activity in hazy areas is decreased. *Copyright © 1998 IFAC*

Keywords: sway mechanism, hierarchical networks, hazy areas, aed theory

1. THE TASKS OF MATHEMATICS AND CYBERNETICS IN THE MAINTAINANCE OF NEW TIMES CONSTRUCTING

Nowadays the main directions of cybernetics are design, control and learning - the stages of sway task in any world strata: natural (physical, chemical, biological), demographical, engineering (industry, service, conveyance, monetary mechanisms), and knowledge (languages, art, science, design and learning).

Named strata (ordered in agreement with history of their arising) have diverse characteristics in concrete States whose statutes (symbol images) must connect their history with sway strategies in their space. The States exchange diverse details of their own constructions on all strata and these interactions are the base of unions of States. Till now the process of world changing (new strata arising, States and States unions constructing) is realized without actual understanding of its laws and since that with hard errors within the States and in their connections.

In order to correct that, cybernetics must define new world (Kile, 1995, Chestnut, 1995), and create convenient means of its design, control and learning which are constructed as multilayer (hierarchical) networks of States and unions of States (Groumpos, 1995, Groumpos et. al. 1997).

Habitual (onelevel) mathematical theories are incohered with laws of hierarchical systems and they can not meet all practical requirements of design&control&learning process.

Since that mathematical maintainance of new time constructing with hierarchical networks (with advanced technologies of cybernetics) must be realized by new mechanism of mathematics. It is the mechanism of hierarchical mathematics (aed theory) which continues the theories of hierarchical multilevel systems and general systems by M.Mesarovic and Y.Takahara (Mesarovic and Takahara 1970, 1975) and has now its own symbol constructions convenient in sway practice [Novikava et. al., 1995, 1997, Groumpos et al, 1997].

2. THE IMAGES OF LARGE SCALE SYSTEMS IN HIERARCHICAL MATHEMATICS.

Aed theory (A^{λ} mathematics) has now two its main hierarchical symbol images ${}^{\times}\alpha^{\lambda}$ and ${}^{+}\alpha^{\lambda}$ which answer to acts of multiplying (learning) and uniting (design). They contain the new means of control and connect the strata (directions) of A^{λ}. Aed strata are: Λ,λ–level (time),Γ,γ–statute (law, connection), P,ρ–act (process), Ω, ω – unit (state) Σ,σ – construction (contents), B, β – new time (arising level), A, α – sway (coordinator). Aed statute A^{λ} in current level λ is described by its symbol image ${}^{\times}\alpha^{\lambda}$ in following way:

${}^{\times}\alpha^{\lambda}$:

$$\mathrm{A}^{\lambda} \overset{\gamma}{\underset{\rho}{\longleftrightarrow}} \left\{ {}^{\beta}_{\omega}\underset{\rho}{\overset{\gamma}{\mathrm{A}}}{}^{\lambda}_{\sigma} \right\}^{\lambda \xrightarrow[\rho]{\mathrm{A}^{\lambda}} \beta}$$

$$\Lambda^{\lambda} \overset{\gamma}{\underset{\rho}{\longleftrightarrow}} \left\{ {}^{\beta}_{\omega}\underset{\rho}{\overset{\gamma}{\Lambda}}{}^{\lambda}_{\sigma} \right\}^{\lambda \xrightarrow[\rho]{\Lambda^{\lambda}} \beta}$$

$$\Gamma^{\lambda} \overset{\gamma}{\underset{\rho}{\longleftrightarrow}} \left\{ {}^{\beta}_{\omega}\underset{\rho}{\overset{\gamma}{\Gamma}}{}^{\lambda}_{\sigma} \right\}^{\lambda \xrightarrow[\rho]{\Gamma^{\lambda}} \beta}$$

$$\mathrm{P}^{\lambda} \overset{\gamma}{\underset{\rho}{\longleftrightarrow}} \left\{ {}^{\beta}_{\omega}\underset{\rho}{\overset{\gamma}{\mathrm{P}}}{}^{\lambda}_{\sigma} \right\}^{\lambda \xrightarrow[\rho]{\mathrm{P}^{\lambda}} \beta}$$

$$\Omega^{\lambda} \overset{\gamma}{\underset{\rho}{\longleftrightarrow}} \left\{ {}^{\beta}_{\omega}\underset{\rho}{\overset{\gamma}{\Omega}}{}^{\lambda}_{\sigma} \right\}^{\lambda \xrightarrow[\rho]{\Omega^{\lambda}} \beta}$$

$$\Sigma^{\lambda} \overset{\gamma}{\underset{\rho}{\longleftrightarrow}} \left\{ {}^{\beta}_{\omega}\underset{\rho}{\overset{\gamma}{\Sigma}}{}^{\lambda}_{\sigma} \right\}^{\lambda \xrightarrow[\rho]{\Sigma^{\lambda}} \beta}$$

$$\mathrm{B}^{\lambda} \overset{\gamma}{\underset{\rho}{\longleftrightarrow}} \left\{ {}^{\beta}_{\omega}\underset{\rho}{\overset{\gamma}{\mathrm{B}}}{}^{\lambda}_{\sigma} \right\}^{\lambda \xrightarrow[\rho]{\mathrm{B}^{\lambda}} \beta}$$

$$\mathrm{A}^{\beta} \overset{\gamma}{\underset{\rho}{\longleftrightarrow}} \left\{ {}^{?}_{\omega}\underset{\rho}{\overset{\gamma}{\mathrm{A}}}{}^{\beta}_{\sigma} \right\}^{\beta \xrightarrow[\rho]{\mathrm{A}^{\beta}} ?}$$

In this way all aed strata can renovate its original unit A^{λ}, they have all its signs and abilities. The strata $\Lambda, \Gamma, \mathrm{B}, \mathrm{P}, \Sigma, \Omega$ are strongly connected both by their original·unit A^{λ} and by the details of their

${}^{+}\alpha^{\lambda}$:

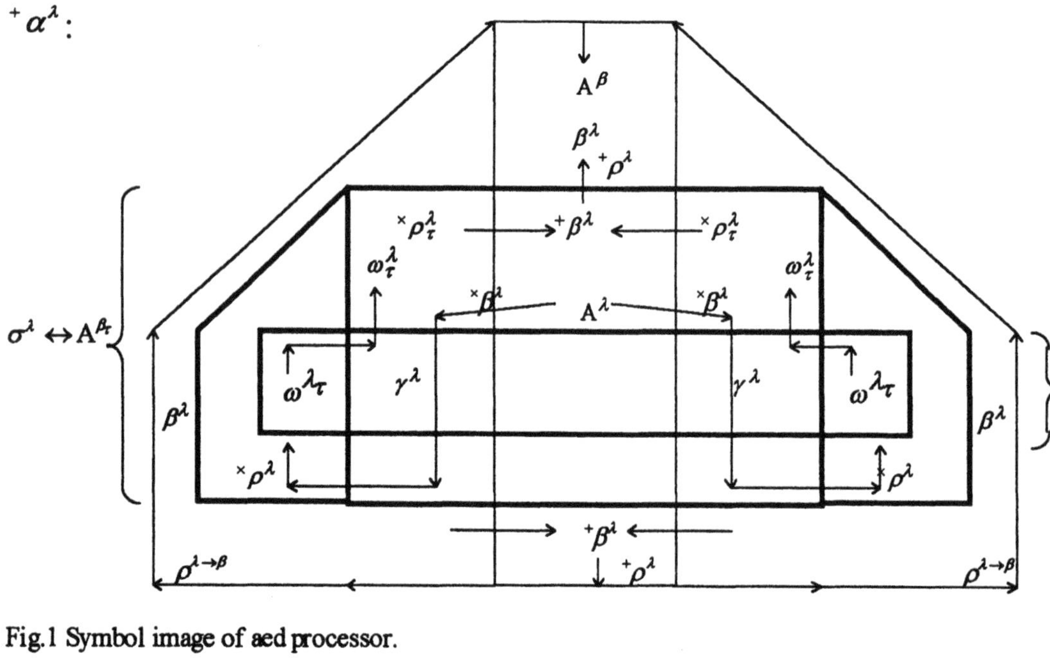

Fig.1 Symbol image of aed processor.

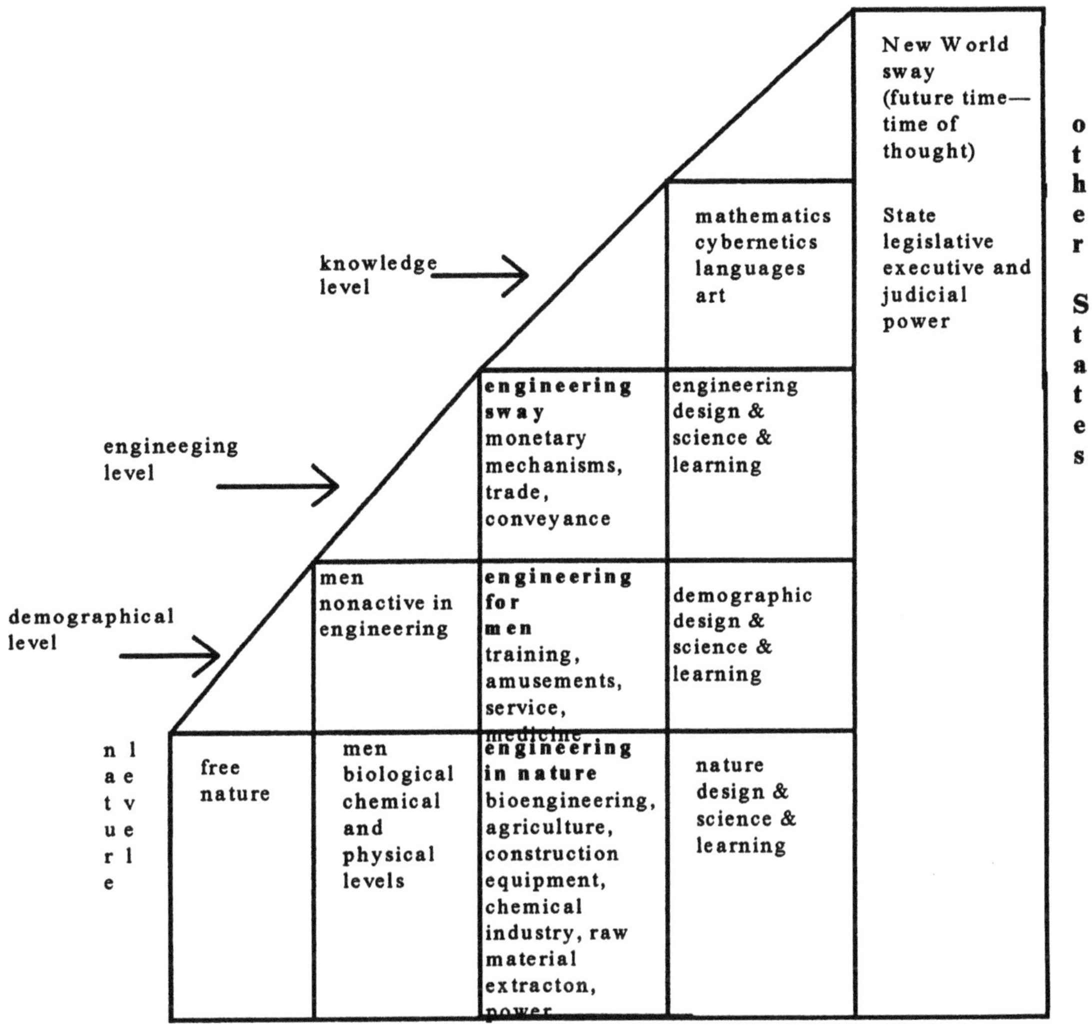

Fig.2 General image of States and unions of States in hierarchical space which contains now the strata of nature (physical, chemical, biological), demographical and engineering levels and strata of knowledge, which realize the sway activity in this space. Future time (time of thought) belongs to the area of hierarchical haze.

own constructions (by their new interactions). Thanks to that all aed strata may be renovated when any stratum is changed. The acts of original unit A^{λ} multiplying and their symbol images uniting lead to the arising of the new time unit A^{β}. It contains hazy symbols in its image (they are signed by the symbol ? and they will be defined in time β) (Novikava et al, 1997). The image $^{+}\alpha^{\lambda}$ (Fig.1) allows to see the acts of multiplying&uniting (learning&design), their bases, aims and connections.

In Fig 1: A^{λ} - original state of aed ($A^{\lambda} \leftrightarrow \omega^{\lambda}$), A^{β} - its leading state in time λ; multiplying act $^{\times}\rho^{\lambda}$ of original order γ^{λ} (original statute of A^{λ}) executes learning process in the units ω^{λ_τ} of construction σ^{λ_τ} which was created in time λ_τ; thanks to that the units ω^{λ_τ} turn into the ordinary units ω_τ^{λ} in the new construction σ^{λ}; uniting act $^{+}\rho^{\lambda}$ in σ^{λ} connects the ordinary units ω_τ^{λ} and creates the new sway $\omega^{\beta} \leftrightarrow A^{\beta}$; in the process ρ^{λ} the image of arising time β^{λ} has two strata - $^{\times}\beta^{\lambda}$ and $^{+}\beta^{\lambda}$ which answer to the stages of this process: $^{\times}\rho^{\lambda}$ and $^{+}\rho^{\lambda}$.

The image $^{\times}\alpha^{\lambda}$ is convenient in the process of mathematical theories connecting, and the image $^{+}\alpha^{\lambda}$ is good in the defining of concrete world strata (physical, chemical, biological, demographycal engineering and knowledge).

The stratum β^{λ} in hierarchical images $^{\times}\alpha^{\lambda}$ and $^{+}\alpha^{\lambda}$ is area of haze (field of aims, arising times). It is changed in time increasing process. The new sway with its new statutes arises in this area and its

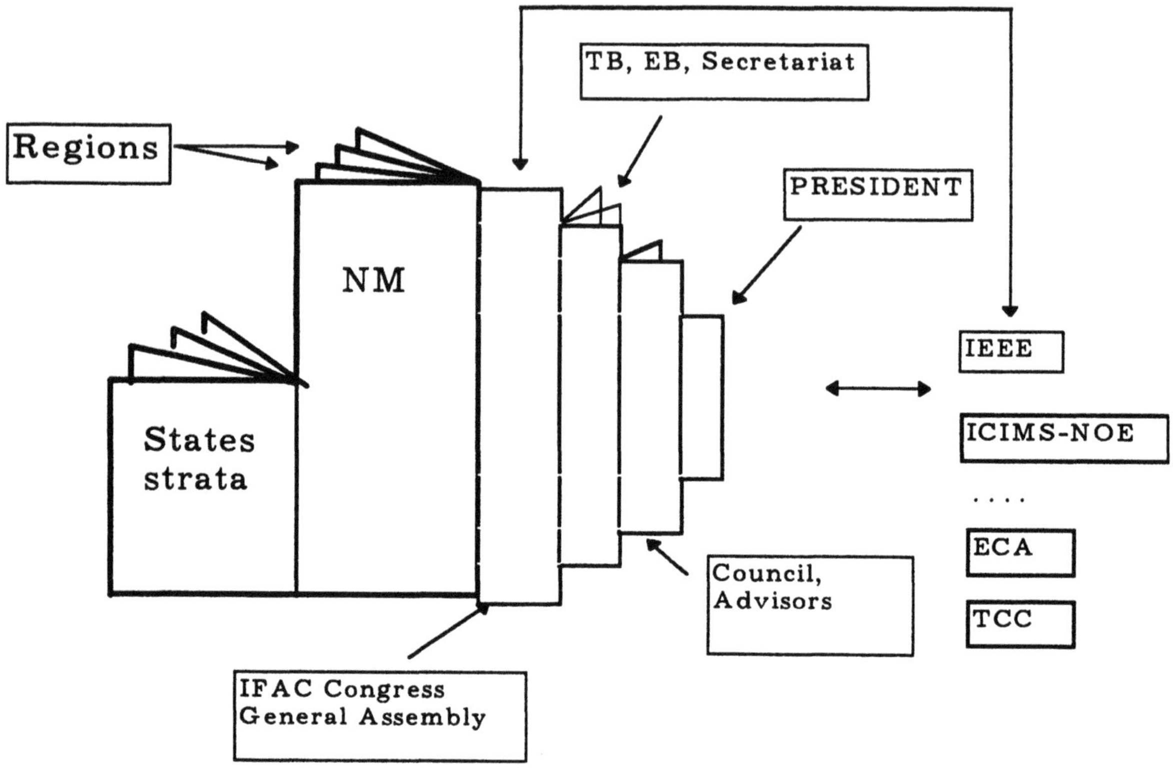

Fig.3 Graphic image of IFAC construction: NM — national (regional) subdivisions. States strata contains now natural, demographical, engineering and knowledge ones.

arising discovers the wider horizons in hierarchical time&space with new haze in it.

Existent world strata may be imaged with graphical means of hierarchical mathematics by Fig.2 which describes the contents of hierarchical time&space and the connections of its strata: lower (natural, demographical and engineering times), highest (knowledge time) and arising (time of thought). Time of thought belongs now to the area of hierarchical haze which cannot have exact images but may be prophesied in the strata of creative work (art, science, design, learning) — in the highest current ranges which contain all other strata of the world and are contained in all existent States. Fig.2 is graphical image of general (key) statute of States and unions of States. At the same time it is the graphical scheme of hierarchical networks which must maintain the activity of sway of States and their unions.

The general lines of thought time arise now in the constructions of leading scientific ranges. Among them the construction of IFAC suggested by H.Chestnut (Chestnut, 1995) as the instance of lucky coherence of thoughts area with practical strategies in the whole world. The IFAC construction in graphic symbols of hierarchical mathematics is imaged by Fig.3. It is easy to understand that Fig.3 (the mathematical image of IFAC construction) is concrete instance of statute of arising time which can be multiplied in diverse States in order to increase the stability in their internal constructions and in their interactions in agreement with the suggestion of H.Chestnut. The main lines of sway networks considered below are defined with taking this suggestion into account too.

3. SWAY NETWORKS OF STATES AND STATES UNIONS: GENERAL CHARACTERISTICS.

The main signs of the whole hierarchical sway network (swaynet) are defined in the following way (P. Groumpos et.al. 1997).

1. The swaynet must allow to define the world:

to learn its history,
to measure its current state,
to design its new image,
to execute its changing strategy.

on all its strata (levels) - natural (physical, chemical, biological), demographical, engineering (industry, service, conveyance, trade, monetary mechanisms), and knowledge (languages, art, design, science, learning, State power institutes (legislative, executive and justice power)).

2. The construction of swaynet contains two ranges (layers, strata):

the field of States swaynets (ordinary units layer); the ordinary swaynets are constructed with taking into account and connecting the internal constructions of their States defined in the world space strata (including hazy ones), their activity as the units in their environment (their interacts, exchange by diverse strata details), their current key unit; and their new horizons in arising times; the States swaynets are the property of their States and they can have the covert details in their constructions; their diversity allows to realize their exchange (including the exchange by knowledge strata details) and the laws of that exchange are the base of their highest strata creating;

the key unit (key range) of world swaynet; the key (sway) unit has its statute and the means of its activity maintainance in all strata (among them the means of mathematics&cybernetics able to carry out all swaynet tasks); it is constructed by the ordinary units in agreement with the main tasks of the swaynet; it answers for the increasing of wellbeing of all lower strata in the world and stratifies the lower units in the ranges cohered with their contributions in its creating and activity; the sway unit is symbol image of the whole hierarchical space with its arising time, it has the turns hazy in its time, it has direct contacts with all lower strata, maintains their diversity, connects them, changes them by the highest achievements multiplying in their own constructions in agreement with their will, creates the means of its own changing, and discovers the way of new strata creating in the world space.

3. The strategy (technology) of swaynet has two main stages:

uniting (design); design technology activates the exchange in the world strata space, unites existent statutes and interacts laws of ordinary swaynets, defines new statute of the whole swaynet, creates its new key unit with its highest strategy in hazy areas including both the means of this strategy executing by the sway unit and the ways of its realizing in all lower strata which must increase their own might;

multiplying (learning); learning technology defines the ordinary units and all lower strata in the symbols of their current highest statute, changes them by the multiplying of highest achievements in their own constructions in line with their will, magnifies their might to the mark more high than the one of their current key unit, create the new diversity in all world strata, and discovers the way of new key statute design in the field of diverse ordinary units learned and stratified in agreement with their contributions in the process of world level increasing; any learned detail of world swaynet is able to renovate it with the increasing of its abilities and the decreasing of outlay in its activity;

all ordinary units of world swaynet have equal rights in its uniting stage and in multiplying stage they are stratified along the new sway ranges in line with these rights realizing by them;

sway rights in the world swaynet space are realized by the ordinary ranges and the key units in turn: the ordinary ranges have sway rights in the uniting stage, the key units have sway rights in the multiplying stage.

4. The main law of world swaynet:

all strata arise in lower strata multiplying and uniting, create more high strata, and are changed by higher strata activity.

The uniting of the strategy of new world order creating with practice of swaynets constructing is good base for the increasing of wellbeing of concrete States and States unions in agreement with aims of the whole world.

4. CONCLUSION

The new cybernetic means (sway networks) defined above in general lines can maintain the process of new time constructing in the world. Hierarchical mathematics allows to construct exact images of these networks and thereby to decrease the high hazard in both their creation and the work with them.

REFERENCES

Kile F. (1995). Desired peace, *Preprints of the IFAC Conference on Supplementary Ways for Improving International Stability, SWIIS'95*. pp.147-152, Vienna, Austria.

Chestnut H. (1995). Improving International Stability and Maintaining Peace, *Preprints of the IFAC Conference on Supplementary Ways for Improving International Stability, SWIIS'95*. pp.11-14. Vienna, Austria.

Groumpos P. (1995). Advanced Controls for Integrated Manufacturing Systems, *Report for ICIMS-NOE Workshop*. Brussels, Nov. 21st.

Groumpos P., S. Novikava, S. Gancharova, A. Zhybul, K. Miatliuk (1997). Design&Creating of New Engineering Units in Reconstructing Regions, *Preprints of Advanced Summer Institute'97, ICIMS-NOE*. Budapest, Hungary.

Mesarovic M.D., Macko D. and Takahara Y. (1970). *Theory of Hierarchical Multilevel Systems.* Academic Press, New York and London.

Mesarovic M.D. and Takahara Y. (1975). *General Systems Theory: Mathematical Foundation.,* Academic Press, New York and London.

Novikava S., et. al, (1995). State Design: New Way in Exact Sciences. *Preprints of the IFAC Conference on Supplementary Ways for Improving International Stability* - SWIIS'95, p.175-181. Vienna, Austria.

Novikava S., S.Gancharova, A.Zhybul, et al (1997). The Statute of Hierarchical Mathematics and Its Cybernetical Maintainance. *Preprints of Eleventh International Conference on Mathematical and Computer Modelling and Scientific Computing.* p.149. Washington, DC, USA.

AUTHOR INDEX

Areshka, T. 85
Astrowskaya, A. 79, 85

Bahutski, A. 79
Barkalin, V. 79
Bludaw, A. 85
Bludaw, Y. 85
Buka, P. 79, 91
Burawkin, A. 85, 91

Caramihai, S.I. 15
Curaj, A. 15

Danichaw, A. 85
Dascalu, M. 63
Dobrescu, R. 21
Dobrescu, S. 21
Dumitrache, I. 15

Filip, F.G. 115
Franti, E. 63

Gancharova, S. 79, 85
Groumpos, P.P. 115

Haber, G. 107
Hersh, M.A. 35

Ionescu, V. 69
Ioska, S. 85

Kalaidjieva, M. 103
Kile, F. 1
Kile, F.O. 115
Kuzminich, I. 91

Levkov, S. 73

Makarenko, A. 73
McKibbin, W.J. 107
Miatliuk, K. 79
Mironychev, A. 29

Neck, R. 107
Novikava, S. 57, 115
Nowik, W. 79, 91

Osipava, H. 85

Paciupa, Yu. 91
Padyninogina, I. 85
Pastravanu, O. 97
Pickl, S. 41

Richardson, J.G. 9

Sarokina, D. 85
Scheffran, J. 41
Shalamanov, V.M. 45
Simion, G. 53
Stanescu, A.M. 15
Stefan, G. 63

Tagarev, T.D. 45
Terahaw, Y. 85
Turovich, A. 85

Voicu, M. 97

Yuhnevich, I. 85

Zezula, A. 85
Zhybul, A. 57, 85
Zianevich, W. 79, 85

www.ingramcontent.com/pod-product-compliance
Ingram Content Group UK Ltd.
Pitfield, Milton Keynes, MK11 3LW, UK
UKHW052302280926
15471UKWH00006B/1057

9 780080 432311